Editor: David Donald

Copyright © 1998 Aerospace Publishing Ltd
Copyright © 1998 Marshall Cavendish

Some of this material has previously appeared in the
Marshall Cavendish partwork 'In Combat'.

This edition first published in 1998 for Grange Books
An imprint of Grange Books plc
The Grange
Kingsnorth Industrial Estate
Hoo, nr Rochester
Kent ME3 9ND

All rights reserved. No part of this publication may be reproduced, stored in a retrieval
system or transmitted, in any form or by any means, electronic, mechanical, photocopying,
recording or otherwise, without the prior written permission of the copyright holder.

ISBN: 1-84013-082-2

Editorial and design by
Brown Packaging Books Ltd
Bradley's Close
74–77 White Lion Street
London N1 9PF

Printed in The Czech Republic

Contents

M60: Destroyer from Detroit 5
Tank v Tank 9
Chieftain! 13
Advance to Contact 17
Merkava: Chariot of Fire 21
Counter-Mobility 25
M1 Abrams: Gulf War Victor 29
Mobile Defence 33
Storm the Line with Challenger 37
Working with Tanks 41
Blast them with the MLRs 45
Ambushing Tanks 49
The BMD and the Blue Berets 53
Tank Hunting 57
T-55: The People's Tank 61
Killing Tanks 65
On Guard with Rapier 69
Counter-Penetration: Plugging the Gap 73
Ratel: Bush Fighter 77
Riding into Battle 81
Cascavel: Armoured Car from Brazil 85
Camouflaging your Vehicle 89
Marauding with the Marder 93
Mobile Operations 97
Panzer Recce!: The Spähpanzer Luchs 101
Eland in Action 105
Break on through with the 432 109
Taking the High Ground 113
Hitting the Beach with the AAV7 117
The Covering Force 121
Racing to War with the Warrior 125
Armoured Reconnaissance 129
Dragoon 133
ZSU-23-4: The Deadly Zoo 137
Saxon the Slayer 141
On the Line with the M109 145
Scouting with the Saladin 149
Battling with the BMP 153
AMX-10P: Armoured Personnel Carrier 157
On the Border with the Buffel 161
M113 in Action 165
Commando! 169
The Bradley in Action 173
Up Front with the VAB 177
MOWAG: Armoured Personnel Carrier 181
Forward with the Spartan 185
Index 190

MECHANISED COMBAT

M60

The M60 Main Battle Tank was the mainstay of US armoured forces for two decades: from the 1960s until it was replaced by the M1 Abrams in the 1980s. It is still in service with the Marines.

DESTROYER FROM DETROIT

American armour has had a chequered history since World War II. Although two US armoured divisions, two mechanised divisions and two armoured cavalry regiments for many years provided an important part of NATO's military strength in Western Europe, it should not be assumed that the 5,000 MBTs used have been of a universally high standard.

The first post-war tank, the M47, was not a success and, although it was exported or donated to friendly governments around the world, it was never fully adopted by the United States itself. The M48, first produced in 1952, was an improvement, but its 90-mm gun was soon recognised as inadequate.

In 1960 the British L7 105-mm gun, built domestically under licence, was fitted to the M48, and by this simple expedient the M60 was born. Manufacture of the M60 proceeded smoothly, with over 3,000 models leaving the Detroit Tank Arsenal production line in the first 15 years alone.

Unfortunately development, if not production, was hindered by government insistence that priority be given to the fitting of a turret capable of mounting the huge 152-mm Shillelagh rocket system then being trialled, with ominous lack of success, on the Sheridan light tank.

M60A2 variant

Three hundred new tanks, designated M60A2, were built with the new enlarged turret, whilst a further 243 conventional M60s were converted to the new specification. The 152-mm gun proved to be far too powerful for the chassis, causing acute stabilisation and fire-control problems, but, rather than concede defeat, the Pentagon squandered millions of dollars

The M60A2 was to be the main production variant, but problems with its complex gun/missile armament meant that it was replaced by the M60A3, armed with a more conventional 105-mm gun.

MECHANISED COMBAT

(which would have been far better spent on the development of the M60 itself) in vain attempts to make the M60A2 gun system battle-worthy. When the Shillelagh eventually entered limited service in 1975, it was found to be hopelessly impractical, and was withdrawn almost immediately.

International agreement

In 1963 a tentative agreement was reached between the United States and West Germany to build an advanced tank along radical new guidelines. When national self-interest led to the disintegration of these joint plans in 1970, both countries decided independently to continue development. Unfortunately, in the USA it was a full 10 years before the first M1 Abrams was ready for trials, which meant that the M60 had to soldier on for much longer than had been planned. Although the production of the M60 continued for an uninterrupted span of 25 years, for much of this period tank development funds were either squandered on the M60A2 Shillelagh or poured into the astronomically expensive M1 Abrams project.

As a direct result the M60 was constantly underfunded and cannot be regarded as a latest generation tank, and although until recently it was an important part of the US Army's front-line armoured inventory, it can in no respect be regarded as a match for the Russian T-64B or T-80.

From late in 1962 the basic M60 was replaced in production by the more advanced M60A1. Although this was a distinct advance on the original, it was still far from adequate. Basic improvements had to be incorporated in the gun, engine and suspension to give the tank a degree of parity with its NATO contemporaries. These were eventually incorporated into the M60A3 variant, which formed the mainstay of the United States armoured fleet, until the M1 Abrams entered large-scale service.

The turret and hull remain of all-cast construction with the Continental AVDS-1790-2A RISE (Reliability Improved Selected Equipment) 12-cylinder diesel engine, capable of 750 bhp at 2,400 rpm, and a top road speed of 48 km/h (30 mph), situated in a large sloping compartment to the rear. The transmission, co-located with the engine, has one reverse and two forward ranges.

The original torsion-bar suspension, with its six road wheels, idler at the front and drive sprocket at the rear, has been enhanced by the introduction of a tube over the bar-suspension on the first, second and sixth road wheels. In addition, the new, much improved T142 track, with removable pads, greatly improves front-line maintenance as well as cross-country performance.

The main armament still consists of

Inside the M60

The M60 is bigger and heavier than any post-war Soviet tank but carries a smaller calibre armament than any Russian MBT since the T-55. On the other hand, better engineering makes it more reliable than some of its potential opponents and superior fire control helps give it the edge in 'quick draw' situations.

Commander's cupola
The large cupola increases the height of the tank, giving the commander a good view but, in the Israeli Army's opinion, making too big a target. The cupola traverses independently and is armed with a .50 calibre machine gun.

Loader
He has his own hatch to the left of the commander's cupola and an M37 periscope with 360° traverse. Thirteen 105 mm shells are stored in the turret for immediate use, 26 in the forward part of the hull and 21 in the turret bustle.

Driver
The driver has a single piece hatch cover that swings open to the right. Like most tank drivers he can be trapped in his seat if the gun is stuck above his hatch but, unlike most, he has an escape hatch in the hull floor for emergencies.

Frontal armour
The armoured glacis of the M60 is protected by 225 mm (8.8 in) of armour plate, which is comparable to the armour of a T-72 and over twice as well protected as the T-62.

The M60 carries more than twice the armour thickness of the T-62. The 105-mm gun is perfectly capable of knocking out early model T-72s as the Israelis demonstrated in Lebanon during 1982, but it needs improved ammunition to cope with the more heavily armoured T-64.

MECHANISED COMBAT

Ballistic shaped turret
The M60A1 introduced this more angular shaped turret, better able to deflect a shell striking the turret from the front. Armour thickness is 250 mm (9.8 in) at the front and 138 mm (5.4 in) on the sides.

Infra red/white light searchlight
Can project a narrow or broad beam of light and can sharply increase its light intensity for up to 20 seconds.

M68 105 mm gun
This is the British L7 105 mm barrel fixed to drop-block breech mechanism. An experienced crew can fire up to eight aimed rounds a minute, much faster than the 115 mm gun on the Russian T-62.

Gunner
The M60A3 has a laser range finder and solid state computer replacing the optical range finder and mechanical fire control system of the M60A1. Range, target and other data is fed to the computer, which then lays the gun accordingly.

Smoke generation
In addition to the usual smoke created by labouring diesel engines, M60s are now fitted with an engine exhaust system, allowing them to create a smokescreen by spraying fuel on to the engine manifold in a similar manner to Russian vehicles.

Suspension
The torsion bar suspension system consists of rubber tyred road wheels with a drive sprocket at the rear and return track rollers. The M60 can ford 1.2 m (4 ft) without preparation and twice as deep if it is prepared.

the old 105-mm M68 gun, but now with full stabilisation in elevation and traverse to facilitate firing on the move. The turret, which can traverse through 360° in 15 seconds and allow barrel elevation of +20° and depression of −10°, is fitted with an electro-hydraulic control system capable of manual override in an emergency. The barrel is provided with a large and ungainly, but effective, bore evacuator to keep fumes in the turret to a minimum and a thermal sleeve to ensure constant performance under all weather conditions.

Mixed rounds

The average rate of fire is stated to be between six and eight rounds per minute. Sixty-three mixed rounds are carried: 26 to the left and right of the driver, 21 in the turret bustle, three under the gun, and 13 immediately

The M60's lofty height would be an embarrassment on a completely flat plain, but billiard-table terrain is rare. Able to depress its gun 10°, the M60 is better able than the much smaller Russian tanks to take up hull-down positions in undulating ground.

available for ready use. It is easy to see that the effect of an enemy projectile penetrating any part of the turret or forward chassis would be devastating for the entire crew!

A variety of ammunition, effective to a maximum range of 2,000 m (6,560 ft), is available, including APDS-T (Armour-Piercing Discarding Sabot-Tracer), HEAT-T (High-Explosive Anti-Tank-Tracer) and Smoke WP-T (White Phosphorus-Tracer).

MECHANISED COMBAT

Israeli M60s were among the first battle tanks to have been operationally fitted with ERA, or explosive-reactive armour, making their debut in the Lebanon during the 1982 operation 'Peace for Galilee'.

M60s have been thoroughly desert-tested. US Army and Marine Corps tanks operated alongside Egyptian M60s in 'Bright Star' desert exercises, as well as being used extensively in the Gulf War.

A 7.62-mm M73 machine-gun is mounted co-axially with the main armament, whilst limited localised protection is provided by a 0.5-inch M85 machine-gun mounted on the commander's cupola. 5,950 rounds of 7.62-mm and 900 rounds of 0.5-inch ammunition are carried.

Variants

Both an M60 AVLB (Armoured Vehicle-Launched Bridge) and the M728 CEV (Combat Engineer Vehicle) were developed from the basic tank. The AVLB consists of a standard chassis with a hydraulic cylinder assembly, and a scissors bridge capable of carrying any NATO tank and of spanning a width of 18 m (60 ft) replaces the turret.

The M728, which first entered service in 1968, is based on a modified M60A1 chassis. Armed with a 165-mm M135 demolition gun capable of firing an M123A1 HEP (High-Explosive Plastic) round, a 7.62-mm co-axial machine-gun and 0.5-inch machine-gun above the commander's cupola, the M728 is also fitted with an 'A' frame mounted at the front of the hull, a hydraulically-operated dozer blade, and a two-speed winch with a capacity of 11,340 kg (25,000 lb) is mounted on the rear. The M728 is now employed by combat engineer battalions to destroy enemy fortifications, fill in trenches, remove obstacles and build defensive emplacements.

The future

At present the M60 and M60A3 are in service with countries as diverse as Austria, Ethiopia, Iran, Italy, Jordan and, of course, the United States. By the time M60 production ceased in 1987, more than 15,000 M60s had been produced and various updating kits, notably the Teledyne Continental powerpack, will take this ageing MBT well into the next century.

In a series of searing attacks made during the US National Guard Adjutant Generals' Conference held at the beginning of 1987, serious shortcomings were identified in the M60A3 TTS (Tank Thermal Sight) tank, and plans were mooted to introduce an updated M60A4. Mobility deficiencies in the suspension and powerpack resulting in a lack of power, marginal control at high speed, and inefficient stiff transmission were highlighted. The high-vehicle profile, virtually useless cupola, poor NBC protection, minimal armour and fire-control deficiencies were also criticised.

Improvements

A number of improvements were suggested varying from the introduction of appliqué armour, the retrofitting of a new 120-mm main gun, and the installation of the AVDS 1790 diesel engine rated at 1,050 hp. It was estimated that updating would cost $730,000 per tank.

The plan was overtaken by events, however, when the US Army ordered the M1A1 tank. This is replacing earlier versions of the Abrams in front-line units, older M1s thus released are being passed onto the National Guard, where they will replace the M60s.

MECHANISED COMBAT

TANK v TANK

Through the ×10 magnification of your commander's sight you can clearly see the outline of an enemy recce tank. The commander of the enemy tank is using his binoculars, but he is not looking in your direction. The squadron leader has acknowledged your contact report, you are in a good fire position, and the enemy is a sitting duck. Fire!

With a 120-mm rifled gun with a maximum range of over 14,000 m (45,900 ft) firing indirectly, or a maximum effective range of around 3,000 m (9,840 ft) in the direct fire role, tied into a YAG laser, and a highly sophisticated computerised fire-control system symbiotically linked to a separate Thermal Observation and Gunnery System, you cannot just yell "Fire!" and hope to destroy a tank. There is a little bit more to it.

Before setting off for the advance you should get your crew to action stations. The gunner will go through his computer and laser checks to ensure the correct functioning of the system (or Improved Fire Control System – IFCS). While he does this, the loader will load the machine-gun and half-load the main armament.

Challenger, and its predecessor Chieftain, has two-piece ammunition; the charge and the round are separate. There are many reasons for this, but there are two main ones: if the tank is hit it is less likely to brew up; and it is difficult to carry 120-mm brass-cased ammunition around.

Half-loading the gun

Half-loading the gun means inserting the round in the breech, but not putting a charge behind it. The gun is quite safe, even if the gunner should accidentally pull the firing trigger: with no charge the gun will not fire. However, the round will plug what otherwise would be a 120-mm hole in the NBC overpressure system.

So the computer is on and working, the machine-gun loaded and made safe, and the main armament is half-loaded. If the target is going to be destroyed, you, as the commander, have to get the crew working towards one aim – killing the enemy. To do this they need to know what the aim is, what you want them to do and when, and what round to fire at it. Therefore, you need to give a fire order.

But what are your choices? What type of rounds do you have, and which one should you choose? You will have two main anti-tank rounds: Fin and HESH – armour-piercing discarding sabot fin-stabilised and high-explosive squash head. The normal Challenger load is 64 rounds, the bulk of them being Fin.

Fin is the latest anti-tank round. It works on a very simple principle. If you take a long, thin and dense piece of metal, put fins on it to give it accuracy in flight (like a dart) and make it

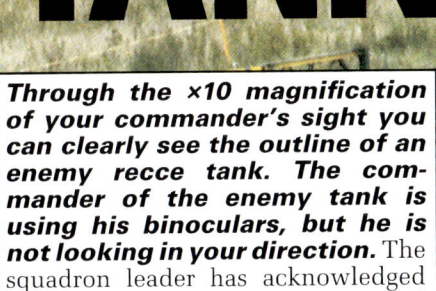
A Chieftain tank fires its 120-mm rifled gun. The main round used against enemy tanks is known as APDSFS, or Fin for short. A dart-shaped piece of very hard metal, it will penetrate the armour plate of most tanks.

Challenger tanks have very gently sloping armour on the hull and turret front. This maximises the thickness of armour an enemy shell must penetrate to destroy the tank.

MECHANISED COMBAT

FIN, TANK, ON!

The sudden appearance of an enemy tank in your sights springs the crew into a well-rehearsed drill; everyone knows what to do and how to do it. Despite modern computers and sophisticated electronics, it is still down to the ability of the crew to operate as a single, highly-trained and professional team if they are to survive.

1 Scan the arcs

Once in position, the crew will scan their arcs; the commander will take the far distance and the gunner the close-range area. Both of them have identical gun control equipment, known as the gunner's or commander's grip switch and gun controller. By the use of a small joystick controlled by the thumb, the 17-tonne (16.7-ton) Challenger turret can be precisely traversed to sweep the arc. The commander's controls will, however, always override the gunner's. On sighting a target, cry "Target left" or "Target right".

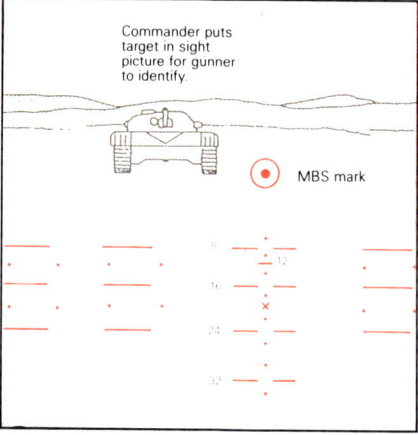

2 "Fin, tank, ON!"

Having sighted his target, the commander will lay the gun in the direction of the target so that the gunner can identify it. Both the commander's and gunner's sights are linked to the gun so that wherever it points their sights will follow. The commander will issue the fire order "Fin, tank, ON". 'Fin' tells the gunner and the loader which ammo type to use; 'tank' tells the gunner what to look for, and 'ON' tells him that the target is in sight and that he has passed over control of the gun.

3 "Loaded!"

The gunner has spotted the target and correctly identified it. He wil' shout "ON", indicating to the commander that not only has he identified the target but also that he has control of the gun. If he does not identify the target, he will shout "Target not identified", in which case the commander will fine-lay the gun onto the exact target. At the same time, the loader will be loading the two-piece 120-mm ammo as fast as he can. Once the gun is loaded he will pick up the next round to be fired, ensure that the breech is closed and that the turret safety switch is on live, and shout "Loaded!"

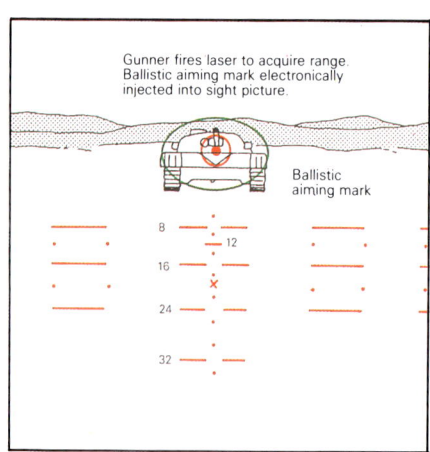

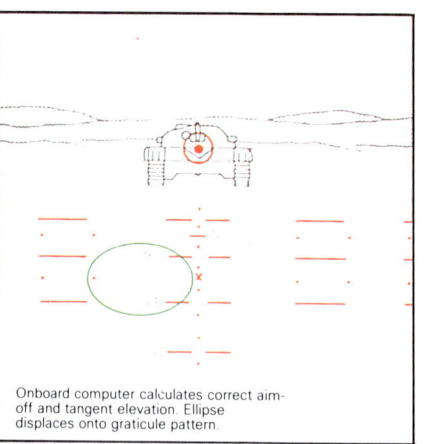

4 "Fire!"

The commander will now check that the gunner has laid on the right target and selected the right ammunition type on the computer, that the gun is ready to fire, and that the firing circuits are live. He will have supervised the loader to make sure that the correct round and charge were loaded. When he's satisfied he will shout the command "Fire!" This does not tell the gunner to fire the gun straight away, but to begin the firing sequence.

5 "Lasing!"

The gunner will fire the laser to get the range to the target, shouting "Lasing!" as he does so. Less than a second later the ballistic aiming mark (an ellipse) will be electronically injected into the sight picture. The computer will calculate the correct elevation and aim off for the gun. The ballistic aiming mark will shift from around the target onto the engraved graticule pattern. A second later, the gun will automatically drive up so that the ellipse is once again around the target. When it is, the gunner will fire, shouting "Firing now!"

6 "Target!"

There will be a huge explosion and a blinding flash as the gun fires. The tank will rock slightly. Both gunner and commander observe the fall of shot. The chances are that, if the equipment is functioning correctly and the proper drills were observed, the round will hit the target. A direct hit using a fin round will result in a blinding white flash. The gunner will shout "Target!", indicating a hit. The commander will reply "Target stop", confirming the hit and finishing the engagement sequence. The loader will still load the next round to ensure that the gun is always ready.

MECHANISED COMBAT

7 "Target stop"

If the preceding engagement sequences have been followed properly and the equipment is functioning correctly, then there is an almost 100 per cent chance that the round will hit the target. That is not to say that the target will be destroyed: if, for instance, the target was moving, the first hit may give an M-kill. It will then be necessary to shoot it again to achieve K-kill. Experience from modern wars, and analysis of the ground over which the next war may be fought, has shown that most tank engagements are likely to take place at around 1,000 m (3,280 ft). But current British tanks are designed to engage at over 3,200 m (10,500 ft) and to engage at that range is a waste of ammunition since the chances of hitting are greatly reduced. Also, at very close ranges (less than 400 m [1,300 ft]) there are special techniques for engaging targets. At that range all you have to do is point the gun at the thing and fire: you will hit something. When planning a tank shoot, an ideal position is one in which the commander can observe the targets at longer ranges, for instance 3,000 m (9,840 ft), and get a good fix. When they come into battle range of around 1,500 m (4,920 ft) he can then engage with an almost certain chance of hitting.

travel very fast, the chances are that if it hits a target it will smash straight through it. This is not a very sophisticated system, but is highly effective. Fin travels so fast and is so accurate that your chances of achieving a first-round hit are at least 90 per cent.

HESH is a chemical round, i.e. it relies on chemical energy, an explosion, to achieve its effect. (By contrast, Fin is a kinetic energy round – it relies on the power of a gun.) The head of a HESH round is made of softened metal that 'pancakes' on impact. The round is detonated, and a shock wave passes through the armour and rips a scab of metal off the inside of the target. This piece of razor-edged metal will then fly around the target at high speed.

Because it flies much slower than Fin, HESH is not accurate. But because it does not rely on the energy of a gun, but on its own explosive power, its lethality is not dependent on range – it is as lethal at 500 m (1,640 ft) as it is at 5,000 m (16,400 ft). It is used mainly as a secondary anti-tank round, although against soft-skinned targets such as APCs, the use of HESH would make more sense.

Fin loaded

You have a Fin round half-loaded and the computer is on and running. When you spot a target, you will give the fire order "Fin, tank, on!", which will instruct the crew to start a series of drills that they have practised. The loader will insert a Fin charge into the breech. He will then slam the breech shut, pick up the next round, close the safety guard and shout "Loaded!". The gunner, at the same time, will activate the firing circuits, use his sight to pick up the enemy tank and place the aiming mark on to the very centre of the target.

Once the loader has shouted

Above: Inside the turret of a British Chieftain. Despite the size of the 57-tonne (56-ton) tank, the interior is very cramped.

Left: The layout of a German Leopard 2 shows how modern tanks store their ammunition behind the thickest armour.

"Loaded!" and the gunner has shouted "On!", you can shout "Fire!". The gunner then fires the laser to obtain the range to the target. At the same time as this is displayed on your read-out, the computer will automatically calculate the elevation and aim-off and will drive the gun to the correct position. When it is in place the gunner shouts "Firing now!" and pulls the firing trigger. All you should see through your sight is a blinding white flash on the target as the solid metal core of your Fin round smashes

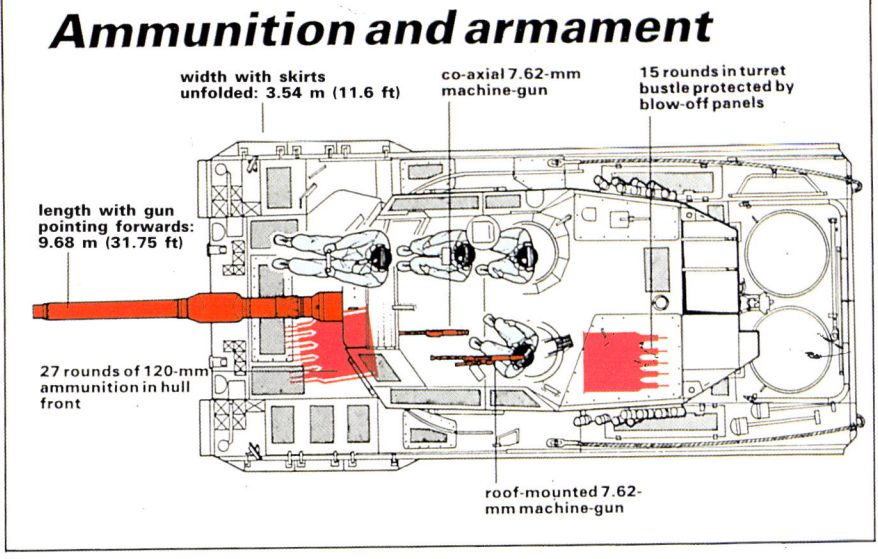

Ammunition and armament

MECHANISED COMBAT

Left: Tank ammunition includes the dart-shaped Fin discarding sabots (far left), seen here in the sabot that strips away as they exit the muzzle. The yellow-topped HESH round contains a charge of plastic explosive that pancakes on to the enemy tank before detonating, blowing scabs of metal into the crew compartment.

Below: The deadly accuracy of modern weapons has made night operations much more common than in World War II, and with night vision aids in widespread use, tank crews now face 24-hour battles.

through the tank armour. At the same time as you see the flash, your gunner, who should also see it, will yell "Target!", indicating that he has seen the hit. From first acquiring the target to this stage should take no more than eight seconds. You have to decide now whether the target is destroyed or merely wounded and needs finishing off.

If you have not destroyed it, shout "Target go on!", and the whole procedure will be repeated. If the tank has been destroyed give the order "Target stop!". In either case the loader will pick up the next round and have the gun ready to fire immediately.

If you miss the target, you must repeat the whole process. The most common reason for missing is an inaccurate initial lay by the gunner. When he was first lining up the target and getting ready to fire the laser, he did not make sure that he was lined up on the centre of the observed target. If he accidentally lined up on a dune in front of the target, for example, the computer will process the calculations for the dune's position. Problems may also arise if you have a dud charge or, perhaps, a freak gust of wind, or one of a number of minute variables that would cause the slightest inaccuracy. In any case, fire again using the same ammunition. If you miss again, fire another round.

If you miss this time it is probably shrewd to withdraw and effect whatever running repairs you can. Your gunner should be well enough trained to fix most problems. If he can't, call the mechanics, who will certainly be able to solve it. Every squadron has its own organic mechanic section, and within this is a gun-fitter who is trained for this task. In the unlikely event that he cannot repair it, the tank will have to go back for more specialist repair work.

However, if your gunner was spot-on, the ammunition was good and there was no freak wind, within eight seconds of first spotting the T-62 you will have fired around £3,000 ($4,650) worth of anti-tank ammo at it (one round), and it should now be destroyed. This is probably the first of many in the forthcoming attack.

This British Challenger has a bank of smoke grenade launchers on the turret front, either side of the main armament. They provide an instant defensive smokescreen around the vehicle, but the latest sighting equipment can see through many chemical smokescreens.

Above: A Challenger tank races around a forest block in Germany. Advances in tank engines make the 51-tonne (50-ton) tanks of today more mobile than some light armour of World War II.

MECHANISED COMBAT

CHIEFTAIN!

The British Army entered World War II in 1939 with few tanks and even fewer ideas of how to use them. Despite the undoubted bravery of their crews, British tanks were outgunned and outperformed in virtually every theatre of war for the next six years.

Only in the last weeks of the war did the British Army receive a first-rate tank when the Centurion Mk 1 entered service. Designers had at last managed to overcome the difficulties of combining a large gun, reliable engine and thick protective armour into one vehicle.

During the 1950s the pressure on designers to improve firepower, mobility and protection increased. All tank designs are the product of com-

Chieftains lumber forward, emitting a characteristic cloud of exhaust smoke. For many years, the Chieftain was the world's most heavily-armed and thickly-armoured Main Battle Tank.

promise between these three factors, but nowhere more than the Chieftain are the dangers of constant modifications apparent.

In many respects initial design studies for the FV4201 Chieftain, or Medium Gun Tank No. 2 as it was first designated, envisaged a low hull, low-slung suspension big brother of the tried and tested Centurion incorporating the existing L7 105-mm gun. However, demands that the gun should be able to outrange the latest Soviet 115-mm T-62 MBT and that the armour should be able to withstand enemy artillery fire from medium calibres increased the weight so much that the main engine already selected proved underpowered.

The initial Chieftain prototype was ready in September 1959 and a series of six prototypes was delivered by the

At the heart of the Chieftain's fighting power is its immensely powerful 120-mm rifled gun, still capable of destroying any known tank at battlefield ranges.

MECHANISED COMBAT

Inside the Chieftain

The Chieftain is one of the most successful post-war Main Battle Tanks. It is slow by modern MBT standards and the reliability of its powerpack is still a problem, but its massive firepower and thick armour plate enable it to survive on the modern battlefield where other 1960s tanks would be hopelessly outmatched.

Leyland L60 engine
The Chieftain's most serious weakness has always been its engine. Supposed to be 750 hp, early models were rated at under 600 hp and it was not until 1979 that BAOR's Chieftains had uprated 720-hp engines. Chieftain is painfully slow compared to Leopard 2 and the M1 Abrams with its infamous whining gas turbine.

Gunner
The main armament is armed with the aid of a laser rangefinder accurate to within 10 m (33 ft) and with a maximum range of 10 km (6 miles). The improved Fire Control System uses a computer to determine the correct elevation and lead required by the target.

Commander
The commander can traverse his cupola independently of the turret and can override the gunner and operate the main armament himself. He can also fire the 7.62-mm machine-gun on his cupola from inside the vehicle.

L11A5 120-mm rifled gun
The best tank gun of its generation, the L11 mainly fires APFSDS-T (Armour Piercing Fin Stabilized Discarding Sabot – or 'fin' for short) and HESH. However, it is over 25 years old and a replacement is overdue; some Chieftains will probably receive the new L30 120-mm gun, which is being retro-fitted to Challenger.

Loader
The Chieftain's rate of fire depends on the loader's ability to slam the heavy shells and their charges into the breech. This is doubly difficult on the move when the breech goes up and down as the stabilization system keeps the gun trained on target while the rest of the tank follows the lie of the land.

end of 1962. Early troop trials soon revealed the tank's many problems, particularly with regard to mobility, and the first full service variants were not accepted until early 1967.

The powerpack
The rear-mounted L60 engine was one of the first multi-fuel powerpacks to enter service. In theory, the multi-fuel concept considerably reduces the resupply problems that have always hounded armoured units, since a multi-fuel engine is supposed to run on a wide variety of fuels. In practice the Leyland L60 proved a complete failure, incapable of producing anything like sufficient power for operational purposes.

The gunner looks through his sights, photographed from the loader's position on the left-hand side of the turret. You load the projectile and charge for the 120-mm gun separately.

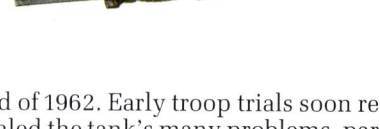

Whereas the American M60's Continental AVDS-1790-2A 12-cylinder diesel could develop 750 bhp at 2,400 rpm and the German Leopard 1's MTU MB 838 Ca M550 10-cylinder multi-fuel pack could develop 830 bhp at 2,200 rpm, only a few of the modified Leyland engines retrofitted to the later-model Chieftains could attain more than 650 bhp. As a result, Chieftain crews still find it necessary to run their engines at maximum revolutions for prolonged periods on exercise, with the result that overheating is a perennial problem.

Unreliable gear-box
Furthermore, the TN12 epicyclic gear-box is prone to breakdown, usually at the most inconvenient time. Sadly, therefore, the Chieftain has gained the unenviable reputation of being one of the most unreliable Main Battle Tanks in world service today.

Chieftain's main armament is as excellent as its engine is bad. The 120-mm L11A5 rifled gun, made at the Royal Ordnance Factory at Nottingham, was undoubtedly the finest tank gun in the world when it first appeared. Technologically, it is inferior to the German 120-mm smoothbore Rheinmetall gun fitted to the Leopard and the US Army's M1A1 Abrams, as well as to the 125-mm smoothbore fitted to the Soviet-built (now Russian-built) T-72 and T-80. In practice, Challengers fitted with L11s out-shot the T-72 during the Gulf War, and did at least as well as the Abrams, particularly at long range.

Elevation of +20 degrees, depression of −10 degrees and traverse of 360 degrees enable the gun to engage any target quickly, and the GEC-Marconi stabilisation system enables the crew to fire while traversing the roughest of country with a good chance of first-round success. A 7.62-mm machine-gun is mounted co-axially with the main armament. Originally the gunner aimed the 120-mm main gun with the aid of a .50-cal machine-gun, but

MECHANISED COMBAT

Searchlight
The box on the left of the turret houses a Marconi white light/infra-red searchlight for target illumination. Modern night vision devices make the use of this kit a disagreeably risky proposition.

Driver
To keep the chassis low, the driver sits in a semi-reclining seat and uses vision periscopes when his hatch is closed. An infra-red periscope can be attached for night driving, but this can be detected by the enemy and an undetectable, passive night sight is being introduced.

Frontal armour
The Chieftain has very thick armour over the hull and turret front but it is conventional armour plate, not 'advanced' armour. Vulnerability to infantry anti-tank weapons forces MBTs to keep at least 500 m (1,640 ft) away from enemy infantry positions, even when fighting in close support of their own infantry.

Side skirts
Often removed for easy access to the wheels, the side skirt provides some measure of protection against the HEAT warheads of infantry anti-tank weapons by detonating them before they actually strike the hull side.

Chieftains are now being fitted with 'Stillbrew'; a horseshoe of extra armour around the turret front. A similar system has been seen on Russian tanks in Afghanistan, and it is a cost-effective way of improving an MBT's protection.

this has now been replaced by the far more effective Barr and Stroud laser rangefinder. A further 7.62-mm GPMG is mounted on the commander's cupola, aimed and fired from within the turret. A six-barrelled smoke discharger is mounted on each side of the turret, and up to 64 rounds of 120-mm ammunition and 6,000 rounds of 7.62-mm ammunition are carried.

A variety of 120-mm separate-loading ammunition, including HESH, APDS, smoke, canister and practice are carried, and APFSDS rounds are carried by later models. The separate-loading ammunition, with its separate projectile and charge, is not only easy to load, but also far more convenient and safer to store than conventional ammunition.

MECHANISED COMBAT

Shock waves

A HESH round, once it hits its target, compresses onto the outer skin. Thereafter shock waves caused by the ensuing explosion cause the inner surface of the target's armour plate to fracture, flake off and fly around inside the vehicle, hitting the crew or, more devastatingly, the stored ammunition.

The more advanced APDS round consists of a sub-calibre projectile with a sabot, or lightly sectioned 'sleeve', fitting the residue of the bore. Once the round is fired the sabot splits and falls away, leaving the projectile to travel at very high speed until it strikes and smashes its way through the enemy target.

Construction

The Chieftain's hull is of traditional construction, with a cast front and welded sides, and thick, removable vertical anti-missile skirts are fitted along both sides of the hull to protect the six road wheels and Horstmann-type suspension. Additional body armour can be fitted to the turret and chassis front to afford added protection against the latest generation of anti-tank weapons. This is remarkably similar to the 'horseshoe' armour seen in the late 1980s on Soviet tanks in Afghanistan.

Although the Chieftain is no longer in service, having been superseded by the far more modern Challenger, some 79 Chieftains remain in reserve. Challenger is a considerably upgraded Chieftain, which entered British Army service in the 1980s. A total of 426 Challengers and 36 of the vastly improved Challenger 2s are currently in service, with a further 350 Challenger 2s scheduled.

Above: By the 1980s the Chieftain had become vulnerable to the latest anti-tank weapons. Chobham armour around the turret was the answer. Its make-up is secret, but probably consists of titanium steel, and ceramic layers. This composite armour also protects the Challenger.

Below: The sun rises over a formation of Chieftains. Once the mainstay of British armoured forces, the Chieftain is now no longer in service, and only 79 are in store, having been replaced by the much superior Challenger 2 in the late 1990s.

MECHANISED COMBAT

ADVANCE TO CONTACT

Above: The M1A2 Abrams is protected by the latest armour, carries a formidably powerful 120-mm gun and can travel at unprecedented speeds across country.

A Challenger tank crosses an anti-tank ditch which has been filled in with fascines. All hatches are closed in case the enemy artillery is zeroed in on the crossing point.

Tanks are for one thing — shock action. Only a tank has the unique combination of firepower, protection and mobility. Although the Israelis launched tank attacks in 1973 with no infantry support and were slaughtered by Egyptian soldiers armed with guided missiles, tanks will still be needed to lead any advance. And advance is the name of the game.

The advance to contact is certainly a well-trialled option. You are out waiting for a fight, and you are going to have one when you find an enemy. The advance can be for one of any number of reasons: to retake lost ground, to force an enemy to withdraw or to re-establish a border. Whatever the reason, the tactics are the same.

The operation is on a brigade scale; in other words, the entire brigade is moving. The three battle groups are spread out, with two leading and one held in reserve. In each battle group there are two tank squadrons of 14 tanks leading, and the third is in reserve with the Warrior company.

Tank formation

A squadron comprises four troops of three tanks and the two headquarters' tanks (the squadron leader's and the squadron 2IC's). The usual formation is two up, the squadron leader travelling with the lead tanks and the 2IC running the rear two. And within each troop the principle is that no tank moves unless someone is covering it, so in a high-threat environment the best way to advance is to leapfrog by troop – one tank moves as the other two cover it. When this tank goes firm, the other two move as it covers them.

When you advance, you do so in bounds. The definition of a bound is 'a tactical feature that may be held if

MECHANISED COMBAT

troop corporal's) advance by leapfrogging, one bound at a time. But even this is not as straightforward as it may seem. How do you get from one bound to the next, and where do you go when you get there?

Movement between bounds must be as fast as possible – a tank on the move is an easy target, and it is very difficult for a moving tank to return accurate fire. So, imagine that you are in your fire position on the back of the bound, and the lead tank has gone firm one bound ahead.

The first thing *not* to do is to drive forward. You always reverse out of a fire position, for two reasons. Firstly, if the enemy has seen you, he will know exactly where you are going to be and will be able to fire at you as you move. Secondly, you will be going over a skyline. This would present a full-size target, moving slowly and probably silhouetted by the sun. Short of painting a big target on the tank, you couldn't help the enemy more.

Move into position

When you pull into the position, the driver should put the vehicle into reverse so that you can pull back immediately. Reverse off the slope and into some cover. If you are on a ridge, get right into the valley, drive along it until you reach the open end of it and follow it round, keeping to the low ground until you come to the bound from which you want to fire.

Then, slowly pull up the slope, with your driver using as low a gear as possible to produce the minimum

Left: An American tank commander prepares to engage long-range targets during desert warfare exercises.

Below: An M60 fires from the short halt. Note the extra height of the commander's cupola.

necessary, and the next tactical feature which enables supporting tanks to give cover.' This translates as a ridge of ground, or even a fold in the ground, which you can drive the tank up and stop so that the earth gives you cover. But, in addition, you must also be able to see over the ridge and, if necessary, shoot over the tanks in front of you.

Ideally, there should be not more than 1,000 m (3,280 ft) between bounds. If you are going to provide cover for a tank to your front, then you have to cover the 1,000 m (3,280 ft) to his position and the same distance again to his front. Usual engagement distance is about 2,000 m (6,560 ft).

The troop of three tanks (the troop leader's, the troop sergeant's and the

MECHANISED COMBAT

amount of smoke, until you can just see over the top of the slope. There is no point in exposing more of the tank than you have to, and at this point all you need to do is to look over the ground.

Use either your commander's sight or, if you are not closed down, binoculars. If the ground is clear, tell your gunner, who has his own sight slightly lower than yours, to tell the driver to advance until he, the gunner, can see the ground as well. As soon as he can see it, you know that if anything pops up, the gun will be above to shoot at it. Scan your arcs and when you think it is clear, radio back and report that you are firm. As soon as you are, the tank that was covering you can move.

That is how you get to your bound, but how do you decide where to go on it? There is a mnemonic for remembering it: CRABMEAL – Cover, Routes, Arcs, Background, Mutual support, Enemy, Air, Landmarks. You want a position that gives you cover from fire. Hiding behind a bush is not cover, but a large thick sand dune can be used. It must have a good route into and out of it, and you must be able to cover the entire arc from it. The back-

Above: M60 tanks fire and manoeuvre like individual riflemen – one moves while the other halts, ready to shoot. Inset: From spotting the target to firing should take no more than eight seconds.

Below: Even with the latest gunnery computers it is difficult to hit the enemy while your own tank is moving. So tanks drive at top speed from cover to cover, kicking up huge dust clouds when fighting in the desert.

MECHANISED COMBAT

ground should disguise your position (easier said than done in the desert), you must be able to cover other tanks from the position and they must be able to cover you. Also, needless to say, the position must face the enemy. You should try to get some cover from the air, but you should not stop by the only tree for 320 km (200 miles) – everybody will be looking at it.

So what happens if there is a target – for instance, an enemy recce vehicle. The obvious answer is to shoot at it. But before you do, get on the radio and give a warning: "Contact, tank, wait out." At the same time, your crew should be getting into action. The loader should be getting the gun ready, the gunner laying the gun and waiting for your order, "Fire!".

Top right: The driver is more than just a stick-puller. On the move he must think about where he will go next and the route he will take, while ensuring as smooth a ride as possible so that the crew can fire accurately.

Right: Even in open terrain experience has shown that tank engagements tend to take place within 1,500 m (4,921 ft). Long-range accuracy is certainly useful, but in tank-versus-tank actions the winner is usually the first tank to fire.

Below: An M1 Abrams waits in a prepared 'hull down' position with only its thickest armour (the turret front) available for the enemy to shoot at. Israeli tanks fighting from similar positions in 1973 inflicted massive losses on Syrian tanks advancing in the open.

MECHANISED COMBAT

MERKAVA

CHARIOT OF FIRE

The Merkava is one of the best protected tanks in the world. Note the sharply angled turret front and gently sloping glacis that maximise armour protection across the frontal arc.

The Israeli Merkava tank is the product of fear. Fear induced by the casualties inflicted on the Israeli tank arm personnel during the 1973 war, and fear that tanks might cease to be available from abroad.

These two statements require explanation. The first, the 1973 casualties, came from several factors. When the Egyptian army crossed the Suez Canal to start the Yom Kippur War of 1973 it was a surprise attack that caught the Israelis off guard. They responded by making rapid, desperate counter-attacks as soon as tanks, with hastily-summoned crews, could be assembled. Often those counter-attacks were made without proper preparation or support from aircraft, artillery or infantry. The Israeli tanks repeatedly rushed the Egyptian forces across wide-open expanses of desert straight into the teeth of waiting anti-armour weapons.

The most effective of the Egyptian weapons were the AT-3 'Sagger' anti-tank missiles that could knock out the Israeli tanks at long range. Nearly every Israeli tank hit meant casualties to the personnel. Most casualties came from the Israeli commanders' habit of heading into battle with turret cupolas open, to improve visibility. Obviously vulnerable to enemy artillery air bursts and small arms fire, many Israeli tank commanders were immobilised this way.

Tank losses

By the end of the Yom Kippur War nearly two-thirds of the Israeli tankies were casualties, and many tanks were out of action for months until they could be recovered and rebuilt (the Israelis never accept that a tank is useless, unless it is a ravaged hulk). But it was the personnel who were the greatest loss to the small nation. Every Israeli tankie takes a lot of training to produce the levels of proficiency expected. Every casualty, therefore, is a considerable loss, not only to the tank arm but to the Israeli nation as a whole.

Protection

Hence the first fear, that the casualties of 1973 could be repeated. Therefore the immediate requirement for Israeli tanks was protection and survival on the battlefield, both for the tank and its crew. None of the then-current Israeli or foreign tanks met these requirements. The only answer was to produce their own design.

That course would also overcome the fear that tank imports would cease. Ever since Israel became a nation in 1948, tank requirements had been met by imports, originally Shermans and French AMX-13s, then British Centurions, and eventually ex-West German M48s and American M60s. It was conceivable that political considerations might prevent future deliveries. The Israelis were not impressed by the quality of the tanks they captured. Once again the answer was to build in Israel.

Home production

Thus, by 1970, an Israeli tank development programme was under way. The Israelis were aware that developing tanks is a lengthy and expensive business, so they overcame much of the process by simply adapting what they had. American, British and French designs were examined, and many components were incorporated in what was to become the Merkava.

Merkava means 'chariot', and the

21

MECHANISED COMBAT

A Merkava with all hatches closed. Israeli tank commanders have always preferred to fight with their head out of the turret, but the risk from snipers and RPG attacks in the Lebanon forced them to button up.

Merkava was to become Israel's armoured chariot. It is an unusual design, tailored to meet the exacting demands of the Israeli tank arm, and heavily protected. This means plenty of thick armour, and the positioning of the engine pack in front of the hull, instead of the more conventional rear. So, even the engine adds protection to the four-man crew. The forward-mounted engine also means that the turret is well to the rear. Sloping armour (which serves to increase armour thickness) covers the long front hull and the turret.

Mobility

Protection impedes mobility. The Merkava is powered by a modified American engine, normally used in the M60 tank series. Another lift from the M60 is the transmission. The road wheels are very similar to those used on the Centurion, but the Israelis have adapted them so that they now have twice their former service life.

The use of items not intended to power such a heavy vehicle as the Merkava (which weighs in for combat at around 60 tonnes (59 tons) means that it is slow, but to the Israelis that means little. What matters is that it can take more combat punishment than most other tanks. Turret and weapon drives are rapid to provide quick responses to fleeting targets.

Weaponry

Firepower is, at present, provided by a 105-mm gun, the same as that used by the Israeli Centurions and M60s. Originally this was a British weapon, the L7, produced in the States as the M68 and now made in Israel as well. In the latest variant the 105-mm gun has been replaced by a more powerful 120-mm weapon.

The Merkava uses an Israeli-designed fire-control system, known as Matador, coupled to a laser rangefinder. Night fighting equipment is limited, but is scheduled for improvement. Here the forward-placed engine could be a problem; the heat produced could fog the thermal-imaging, or infra-red vision and sighting systems. This problem has been partly overcome by insulating the engine compartment, and directing exhaust fumes to one side of the vehicle. More insulation is provided by using the space between two layers of front hull armour for fuel storage.

Protection does not mean armour alone. The Merkava must keep its gun in action for as long as possible, so it carries a lot of ammunition. Ninety-

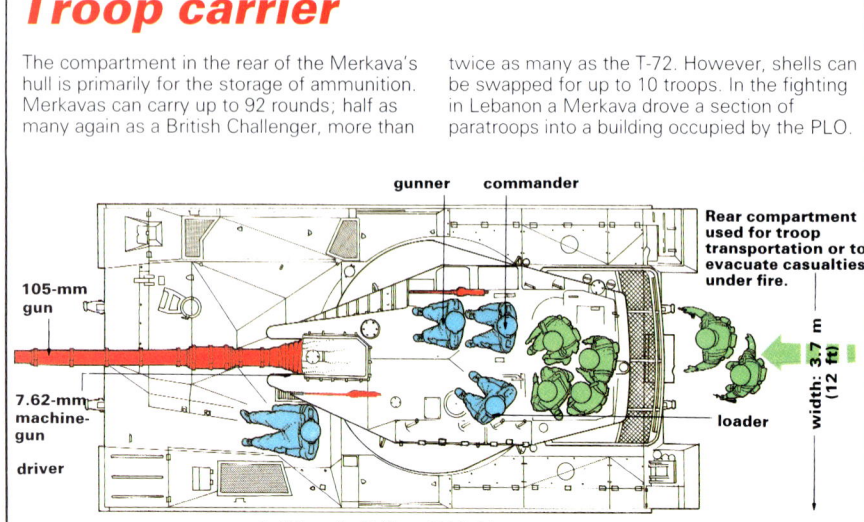

M68 105-mm gun
The Merkava's main armament is the Israeli-produced version of the M68 105-mm rifled gun also used by the IDF's Centurions and M60s. The Merkava has been designed to accept a 120-mm gun, which may well appear on the Merkava Mk 3.

Teledyne Continental diesel
The Merkava's engine is frontally mounted, which provides more protection for the crew. Developing 900 hp, it is a more powerful version of the engine used in Israel's M60s and the transmission is similar. The whole engine can be replaced in the field in about an hour.

Frontal protection
The Merkava has two layers of frontal armour with the space in between filled with diesel fuel. This spacing reduces the effect of the shaped-charge warheads of infantry anti-tank rounds and HEAT shells from enemy armour.

Driver
Unlike many tank drivers who are marooned in their own tiny compartment, the Merkava driver can get into the crew compartment by folding the back of his seat down.

Troop carrier

The compartment in the rear of the Merkava's hull is primarily for the storage of ammunition. Merkavas can carry up to 92 rounds; half as many again as a British Challenger, more than twice as many as the T-72. However, shells can be swapped for up to 10 troops. In the fighting in Lebanon a Merkava drove a section of paratroops into a building occupied by the PLO.

MECHANISED COMBAT

Inside the Merkava

The Merkava is one of the few tanks in the world capable of surviving a hit from another tank's main armament. Several Merkavas were hit in 1982, but there were no crew casualties. By contrast, the T-72s used by Syria burned easily; probably because a penetrative hit cannot fail to ignite the cartridges in the automatic loading system. The Merkava has no ammunition in the turret: eight rounds are stored below the turret ring for ready use and the remainder are in the rear of the hull.

Commander
The Israelis disliked the high cupolas of American tanks and removed them from their M60s. The Merkava has no cupola, but the hatch can be raised to allow the commander to see without using periscopes while still enjoying overhead protection.

Gunner
Using a digital fire control system and laser rangefinder, the Israeli Merkava gunners achieved impressive first-round accuracy against Syrian T-72s in June 1982.

Turret bustle
The Merkava's radios and hydraulic systems are mounted in the turret bustle. Chains have been fitted to the turret bustle to detonate HEAT rounds before they strike the turret ring.

NBC pack
The right-hand hatch on the hull rear provides access to the Merkava's NBC systems.

Telephone
Like some NATO tanks, the Merkava has a telephone for infantry to talk to the crew. It takes considerable confidence to stand next to a 60-tonne (59-ton) armoured monster which may elect to change firing position just as you pick up the receiver.

two rounds of 105-mm ammunition can be carried, plus as many as 10,000 rounds of 7.62-mm ammunition for the Merkava's three machine-guns. The bulk of this ammunition is carried at the rear, behind a two-part hatch, thereby enabling easy access for loading.

Action

The Merkava has proved itself to be a formidable fighting machine. It first saw action in Lebanon, in 1982, where it easily knocked out Syrian T-72s. Its tough protection enabled it to fight at

The Merkava is a huge vehicle; the turret roof is 2.7 m (9 ft) above the ground. However, it can depress its gun by 8.5°, enabling it to take up hull-down positions and expose only a tiny proportion of its bulk.

23

MECHANISED COMBAT

The Merkava first saw action in 1982 when Israel invaded Lebanon, attacking Syrian forces in the Bekaa valley and bombing Beirut. Supported by TOW-firing helicopter gunships, Merkavas had little trouble with Arab T-72s.

close quarters, in built-up areas, where tank-killer squads would have immobilised most other tanks without difficulty.

The one limitation at present is the Merkava's limited night combat capability. The crew can use infra-red driving lights, but target acquisition can be problematic in the dark. Therefore, the Merkava carries 60-mm mortar firing flares to illuminate potential targets.

The Merkava is still being developed, and its potential as a battle tank is enormous. The Mk 2 has many detailed improvements over the Mk 1. Combat experience has added a mantlet of chains around the turret base to protect the turret ring from rocket grenades, and the co-axial machine-gun port now has extra armour. Some Merkavas carry 12.7-mm Browning heavy machine-guns on the turret for air defence, or fighting in urban areas. The commander's cupola can now be lifted to a flat interim position that enables a 'head-out' vision capability, while still providing overhead cover. It is anticipated that flame-proof armour will soon be added.

Improved version

The Merkava Mk 3 entered service in 1988. Equipped with a new 120-mm smoothbore gun, improved armour and suspension, new fire-control equipment and a 1500-hp engine, the Mk 3 lifts the Merkava into the class of the M1 Abrams or the Challenger 2 Main Battle Tanks.

The Merkava is not an all-purpose battle tank for all men. It has been tailored so closely to Israeli combat requirements that it would be ponderous on many other battlefields. Their one problem is that there are too few of them. Merkavas are expensive for a small nation to produce, and production is slow. So the older Centurions and M60s will have to remain in service, with Merkavas, for some years.

Merkavas are often seen with several machine-guns mounted on the turret hatches. The primary threat in Lebanon came from Arab foot soldiers with anti-tank rockets, and multiple machine-guns proved more effective than trying to employ the main armament.

MECHANISED COMBAT

COUNTER-MOBILITY

Tanks are the attacker's main weapons. The engineer's job is to slow them down, destroy the momentum of their attack, channel them into armoured killing areas, separate them from their supporting infantry and destroy them using mines. A comprehensive obstacle plan, carefully sited, will enable you to take on the enemy on ground of your own chosing and at a time most favourable to you.

In a defensive battle, the obstacle plan is a crucial element. Obstacles delay or stop the enemy, making defensive fire tasks more effective and disorganising his attack by restricting his ability to manoeuvre. Even minor natural obstacles, improved as appropriate and covered by direct or indirect fire, will reduce the momentum of an attack and, if successive obstacles are sited in depth, eventually halt it.

Anti-tank ditches are a well tried but effective form of obstacle. Such ditches successfully countered massive Syrian tank attacks on the Golan Heights during the 1973 Yom Kippur War. The ideal anti-tank ditch is of rectangular cross-section, at least 1.5 m (5 ft) deep and 3.5 m (11.5 ft) across with the loose soil heaped on the 'home bank' some 2.5 m (8.2 ft) high. Ditches can be dug by any of the mechanical plant found in a field or field support squadron. This includes the combat engineer tractor.

Anti-tank ditches

Anti-tank ditches should be sited so they can be covered by fire and observations; a reverse slope position is ideal. Ideally ditches should be sited in conjunction with natural obstacles, such as a wood, built-up area or area of marshy ground and a minefield. This combination provides a double problem for the enemy, requiring first a crossing followed by a minefield clearance operation – all under fire.

Roadblocks are erected to stop or hinder tanks or other vehicles on road, tracks or other routes. Such obstacles should also always be covered by fire, particularly anti-tank weapons, and preferably sited with mines. They should be sited so as to make bypassing very difficult.

Roadblocks can be pre-planned and, with pre-stocked equipment, erected quickly in times of war with little effort. But it is the ad hoc obstacle that the Sappers will usually have to construct, often racing against

MECHANISED COMBAT

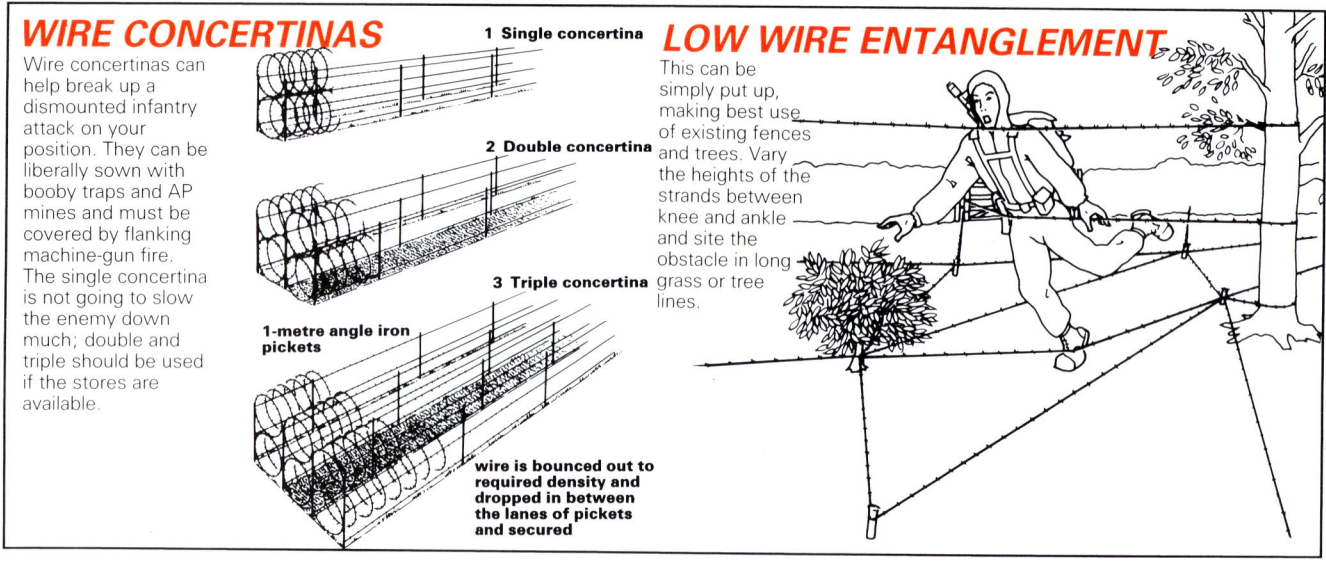

WIRE CONCERTINAS
Wire concertinas can help break up a dismounted infantry attack on your position. They can be liberally sown with booby traps and AP mines and must be covered by flanking machine-gun fire. The single concertina is not going to slow the enemy down much; double and triple should be used if the stores are available.

1 Single concertina
2 Double concertina
3 Triple concertina
1-metre angle iron pickets
wire is bounced out to required density and dropped in between the lanes of pickets and secured

LOW WIRE ENTANGLEMENT
This can be simply put up, making best use of existing fences and trees. Vary the heights of the strands between knee and ankle and site the obstacle in long grass or tree lines.

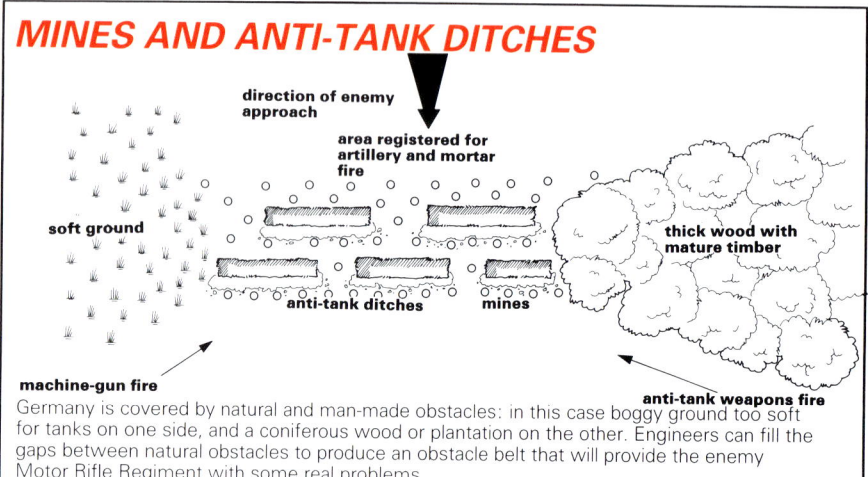

MINES AND ANTI-TANK DITCHES
direction of enemy approach
area registered for artillery and mortar fire
soft ground
thick wood with mature timber
anti-tank ditches
mines
machine-gun fire
anti-tank weapons fire

Germany is covered by natural and man-made obstacles: in this case boggy ground too soft for tanks on one side, and a coniferous wood or plantation on the other. Engineers can fill the gaps between natural obstacles to produce an obstacle belt that will provide the enemy Motor Rifle Regiment with some real problems.

*Soviet troops on exercise in full **NBC** kit during the 1970s. Training with real poison gases, the Soviets fully intended to use their arsenal of chemical weapons against **NATO** defensive positions.*

the clock as the covering force battle is fought.

A typical roadblock would consist of overturned or immobilised juggernauts and – a common continental feature – their trailers. It would be incorporated with anti-tank mines and covered by fire.

Cratering roads

The cratering of minor roads using specially-prepared cratering kits or bulk high explosive is an effective route-denial measure. Sited around a bend, camouflaged where possible and, as always, sown with anti-tank and anti-personnel mines and covered by automatic and anti-tank weapons, a crater will result in the enemy having to deploy dismounted infantry and his own engineers to clear the obstacles before the route is made safe.

Barbed wire is a tried and tested obstacle which plays an important part in holding up the enemy and channelling his movement into pre-planned killing zones. Erection of barbed wire defences is an all-arms task and is primarily carried out by the infantry preparing their defensive positions. However, the Sappers are the experts in all forms of wiring, and it is to them that units turn for advice or guidance when planning or laying out barbed wire defences.

Barbed wire obstacles must, as always, be covered by fire and concealed to have the maximum effect on an enemy assault. Wire must always be outside grenade throwing range of your own positions. Wiring is expensive in manpower and noisy, and so attracts enemy fire and needs careful planning and preparation.

A typical if basic wire obstacle is the single concertina fence. It should be developed into a more complex obstacle, such as a catwire fence, as soon as possible.

Low wire is an extremely effective anti-personnel obstacle. It is comparatively simple to erect, it does not require a great deal of stores because natural fences, bushes and trees can form the basis of the obstacle and, most important, it can be easily and effectively concealed.

Field defences

The Sappers are the experts in the design and construction of field defences and are involved in such tasks where the use of their heavy engineer plant is required.

Field defences are an essential feature of the modern battlefield. Even simple slit trenches provide protection against artillery, mortar and close air support attack and, additionally, against the effects of nuclear radiation. The digging of basic field defences is the responsibility of all units and every soldier is trained in the construction of battle trenches

MECHANISED COMBAT

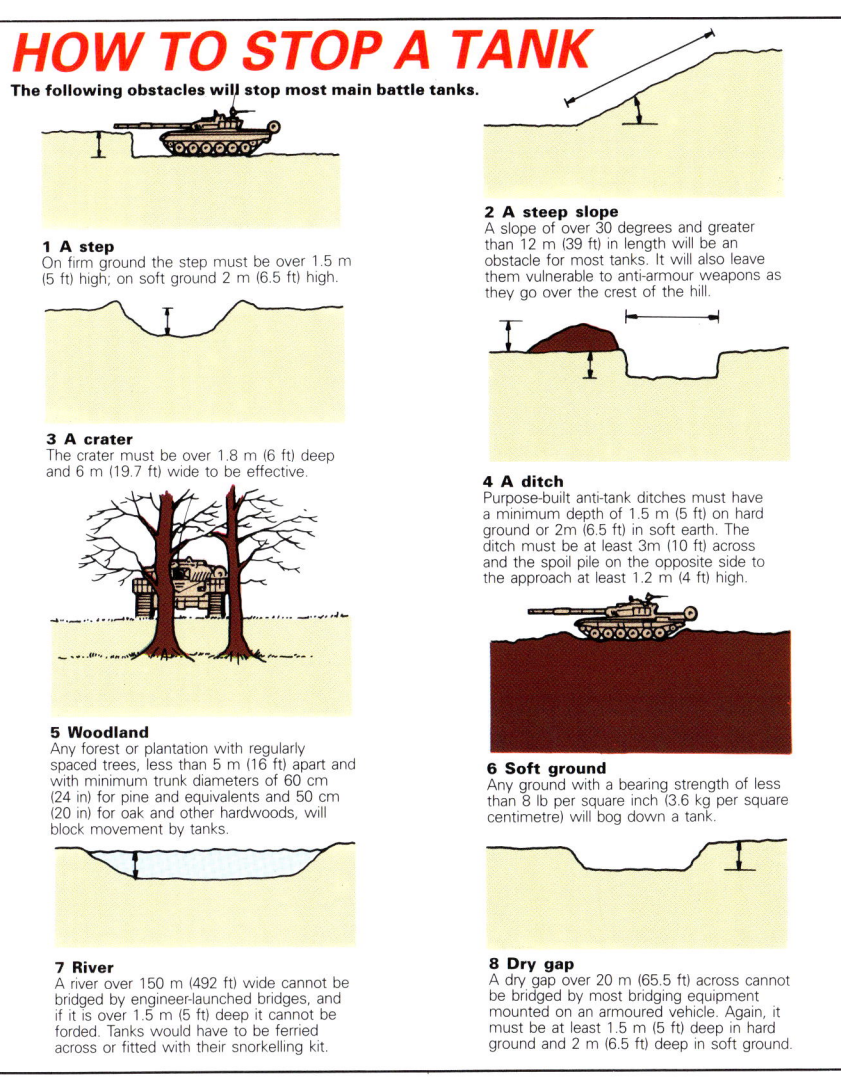

HOW TO STOP A TANK
The following obstacles will stop most main battle tanks.

1 A step
On firm ground the step must be over 1.5 m (5 ft) high; on soft ground 2 m (6.5 ft) high.

2 A steep slope
A slope of over 30 degrees and greater than 12 m (39 ft) in length will be an obstacle for most tanks. It will also leave them vulnerable to anti-armour weapons as they go over the crest of the hill.

3 A crater
The crater must be over 1.8 m (6 ft) deep and 6 m (19.7 ft) wide to be effective.

4 A ditch
Purpose-built anti-tank ditches must have a minimum depth of 1.5 m (5 ft) on hard ground or 2 m (6.5 ft) in soft earth. The ditch must be at least 3 m (10 ft) across and the spoil pile on the opposite side to the approach at least 1.2 m (4 ft) high.

5 Woodland
Any forest or plantation with regularly spaced trees, less than 5 m (16 ft) apart and with minimum trunk diameters of 60 cm (24 in) for pine and equivalents and 50 cm (20 in) for oak and other hardwoods, will block movement by tanks.

6 Soft ground
Any ground with a bearing strength of less than 8 lb per square inch (3.6 kg per square centimetre) will bog down a tank.

7 River
A river over 150 m (492 ft) wide cannot be bridged by engineer-launched bridges, and if it is over 1.5 m (5 ft) deep it cannot be forded. Tanks would have to be ferried across or fitted with their snorkelling kit.

8 Dry gap
A dry gap over 20 m (65.5 ft) across cannot be bridged by most bridging equipment mounted on an armoured vehicle. Again, it must be at least 1.5 m (5 ft) deep in hard ground and 2 m (6.5 ft) deep in soft ground.

with some overhead protection. Specialists in the infantry are skilled in the development of specialised versions of this trench to house the General Purpose Machine Gun (GPMG) or MILAN anti-tank weapon.

All these field defences are usually dug using the basic pick and shovel carried by every infantryman. Explosives can be used to break up the earth and speed up the operation. The infantry's own specialist assault pioneers usually do this.

Specialist know-how
The preparation of larger field defences calls for specialist Sapper know-how, equipment and plant. The most common types of plant which are found in Sapper field units include the Combat Engineer Tractor (CET), which has a large earth-moving bucket and is particularly suitable for digging positions for tanks or self-propelled guns. The Light Mobile Digger (LMD) and the Light Wheeled Tractor (LWT) are also versatile and common forms of plant for digging field defences.

Larger field defences include protective slots for large self-propelled guns and Main Battle Tanks. Because of the Sapper effort involved, the planning of such defences is an engineer responsibility. Large tank slots can conveniently be dug by the CET or an engineer tank (Armoured Vehicle Royal Engineer – AVRE) fitted with a 'dozer blade.

The digging of command posts is an important function of the Sappers.

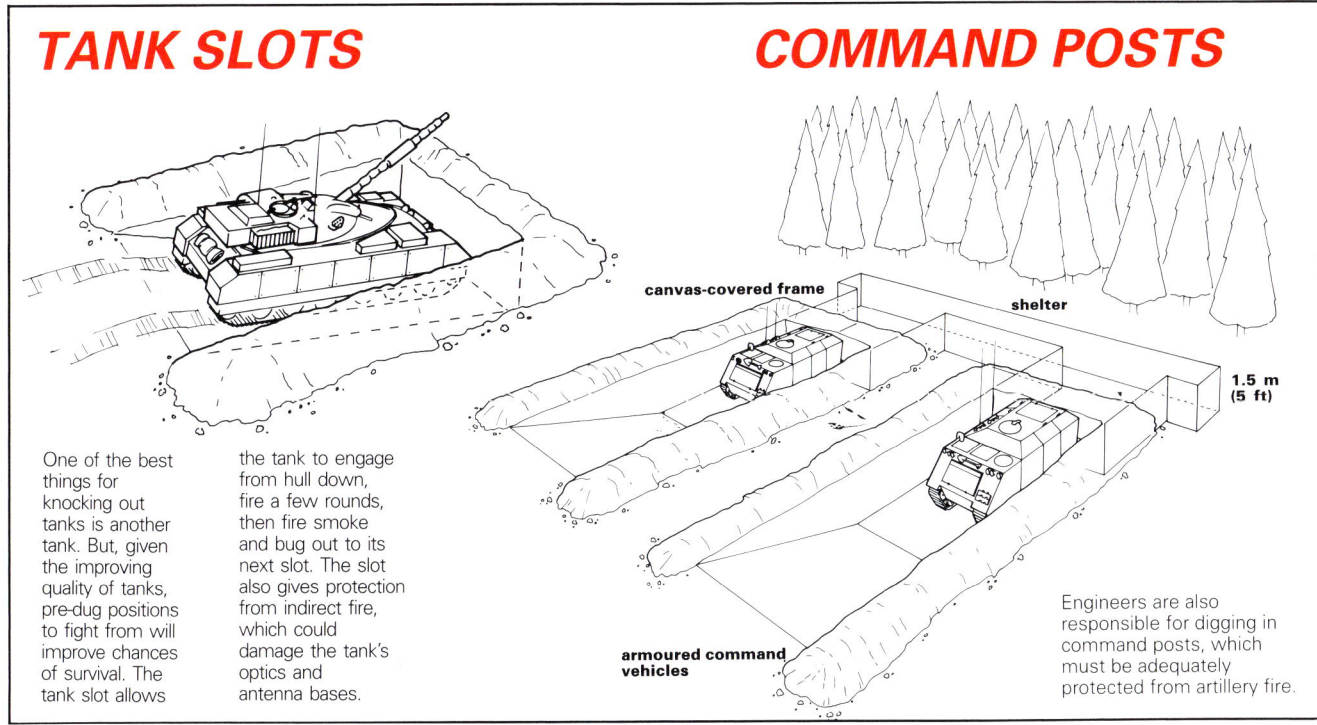

TANK SLOTS

One of the best things for knocking out tanks is another tank. But, given the improving quality of tanks, pre-dug positions to fight from will improve chances of survival. The tank slot allows the tank to engage from hull down, fire a few rounds, then fire smoke and bug out to its next slot. The slot also gives protection from indirect fire, which could damage the tank's optics and antenna bases.

COMMAND POSTS

Engineers are also responsible for digging in command posts, which must be adequately protected from artillery fire.

27

MECHANISED COMBAT

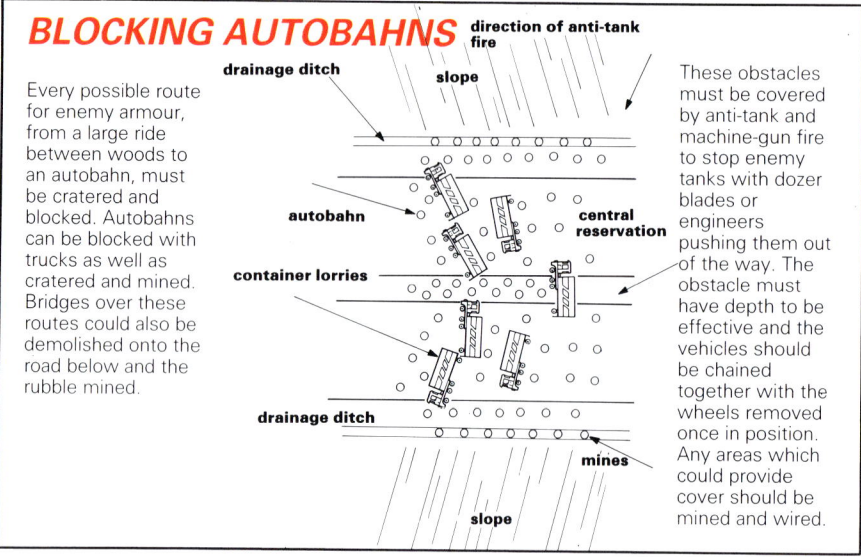

BLOCKING AUTOBAHNS

Every possible route for enemy armour, from a large ride between woods to an autobahn, must be cratered and blocked. Autobahns can be blocked with trucks as well as cratered and mined. Bridges over these routes could also be demolished onto the road below and the rubble mined.

These obstacles must be covered by anti-tank and machine-gun fire to stop enemy tanks with dozer blades or engineers pushing them out of the way. The obstacle must have depth to be effective and the vehicles should be chained together with the wheels removed once in position. Any areas which could provide cover should be mined and wired.

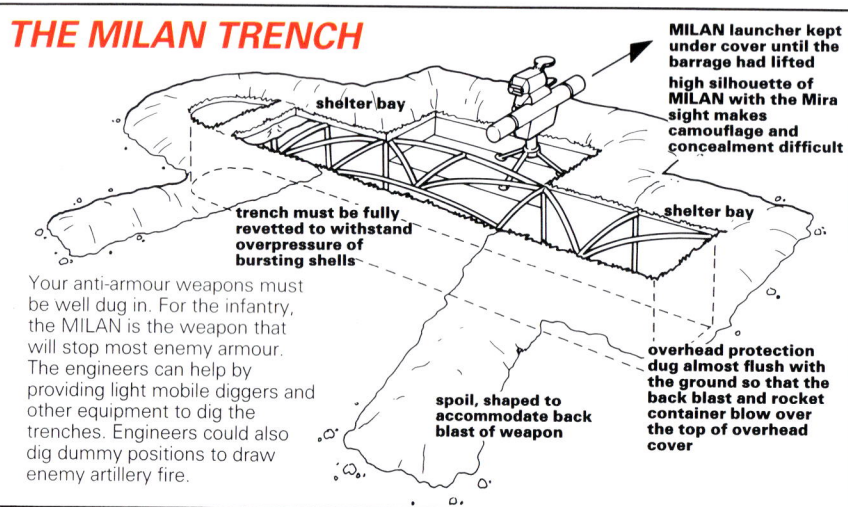

THE MILAN TRENCH

Your anti-armour weapons must be well dug in. For the infantry, the MILAN is the weapon that will stop most enemy armour. The engineers can help by providing light mobile diggers and other equipment to dig the trenches. Engineers could also dig dummy positions to draw enemy artillery fire.

- MILAN launcher kept under cover until the barrage had lifted
- high silhouette of MILAN with the Mira sight makes camouflage and concealment difficult
- trench must be fully revetted to withstand overpressure of bursting shells
- shelter bay
- spoil, shaped to accommodate back blast of weapon
- overhead protection dug almost flush with the ground so that the back blast and rocket container blow over the top of overhead cover

Command posts consist of a field shelter dug in a variety of configurations with armoured command vehicles, usually of the 430 or CVR(T) type. They provide overhead protection from artillery and mortar fire and the radiation and some of the blast effects of nuclear weapons. Similar structures can be used to provide shelter for a medical aid post.

The basic shelter is a tubular frame, covered by a reinforced fabric covering. Dug in, it is further reinforced by overhead protection – packed earth or sandbags. Typical plant that can be used for digging in the shelter of the command post include the ubiquitous CET, the LMD or LWT. A typical dismounted battlegroup command post will require excavation to a depth of 1.35 m (4.4 ft) before erecting the tubular framework and constructing the necessary overhead cover.

When associated with the command vehicles of an armoured formation, the command post is extended appropriately. The vehicles are dug in, in ramped slots backing into the shelter.

The shelter is designed to house a company command post; additional units can be added to house the additional manpower and radios associated with larger headquarters.

In defence, constructing major obstacles, laying minefields, cratering roads, erecting wire defences or large field defences, the Sappers live up to their motto *ubique* (everywhere) and fulfil their role of helping the other combat arms to move and fight.

Ranger anti-personnel mines can be laid by firing them out of these pre-packed tubes fitted to a vehicle. This system can quickly convert an anti-tank minefield into a mixed minefield with little risk to friendly forces and at great speed. The system can be fitted to any vehicle, including this interesting Land Rover half-track.

MECHANISED COMBAT

M1 ABRAMS
Gulf War Victor

The M1 Abrams entered service with the US Army in 1980, when it began to replace the M60 tank in front-line armoured divisions. Compared with the M60, the Abrams is much faster, far more heavily protected, and has vastly superior sighting and fire control equipment.

The nomad camel herder driving his recalcitrant charges across the arid terrain of the Middle East is following a way of life that has hardly changed for a thousand years. But that timeless way of life is about to be rudely interrupted. The camel caravan is in the path of the US Army, which is storming forward with some of the 20th century's most advanced and potent fighting machines.

The desert sand begins to vibrate. Borne on the wind is the unmistakeable sound of tank tracks, a clatter of metal against metal. Then there's something else – not the rumble of low-revving diesels, but a high-pitched whine, the sound of a gas turbine engine spooling at three quarters of maximum power.

American M1 tanks come in fast, six of them in staggered echelon, the lead tank lurching off the rough track and heading straight over the broken ground without any perceptible drop in speed.

The M1's advanced torsion bar suspension simply soaks up the bumps. The American tank's gas turbine engine can churn out 1,500 horsepower, punching this extraordinary

The Abrams made its combat debut in the largest multi-national operation since the Korean War. In the Gulf War against Iraq, some 2,000 M1s formed the backbone of the most powerful armoured force ever assembled.

29

MECHANISED COMBAT

The M1 entered service armed with the long-serving M68 105-mm cannon. Although it lacks penetration in the face of the latest tank armour, the M68 is still a very effective weapon.

Inside the Abrams

The gunner sits with the M1's commander in the right of the turret, with the loader on the left. The gunner has highly sophisticated target acquisition aids at his disposal and, once he has found a target, an equally sophisticated automated fire control system which affords the tank a fearsomely efficient first-round kill rate.

Stabilisation system

The main armament's stabilisation system means the M1 can engage targets accurately even while on the move over broken ground. The gunner merely places his primary sight on the target and uses the laser rangefinder to determine the range. The digital fire control computer then determines the correct elevation and offset angle for a hit and the gunner opens fire. Also feeding automatically into the computer is information on wind

fighting vehicle's 60 tonnes (59 tons) up a 10 per cent slope at over 32 km/h (20 mph). On roads this beast could scorch the autobahn at almost 80 km/h (50 mph). In its latest M1A2 guise, armed with the German-developed Rheinmetall 120-mm smoothbore gun, the Abrams has considerably more fighting power than the contemporary generation of Russian tanks exemplified by the T-64B and T-80.

Traditional design

Yet the latest Russian vehicles incorporate advanced features such as dual-purpose missile-firing armament and anti 'smart' munition electronic warfare defences. The M1 is very much a 'traditional' tank incorporating a wealth of sophisticated electronics, but still a lineal descendant of the M4 Sherman with a tracked chassis and high-velocity gun in a rotating turret.

A tank's firepower is not just a question of the size and number of its weapons. It is a function of its main armament's hitting power, rate of fire and accuracy. The original M1 was armed with a 105-mm gun, but since mid-1985 production has been switched to the M1A1.

The M1A1's 120-mm weapon can fire a range of ammunition, including high explosive anti-tank chemical energy rounds and the M829 kinetic energy round with a slug of ultra-dense depleted uranium as its armour-smashing core penetrator. The M1A1 can carry 40 rounds of 120-mm ammunition, 36 in the rear turret bustle and the rest in a rear hull box.

The M1 Abrams has established new records in the NATO tank gunnery competitions and has proved itself to be an outstanding Main Battle Tank. It has set new standards for armour protection and battlefield mobility, and an improvement programme now under way should further increase the M1's performance.

M68E1 105-mm rifled gun
A modified version of the British L7 gun, the M68 was fitted to the first 3,000 M1 tanks. The Israeli Merkava tank uses the same gun, and it successfully destroyed Syrian T-72 tanks in Lebanon in 1982.

M240 co-axial 7.62-mm machine-gun.
This is another version of the FN MAG, known to the British Army as the GPMG. The M1A1 carries much less ammunition for its 7.62-mm machine-guns, but the feed chute is moved away from the breech of the main armament and the capacity of the spent shell case box increased.

Advanced armour
The new layered armours on tanks like the M1, Leopard and Challenger protect the vehicle from HEAT warheads of anti-tank missiles. Because of this, the Russians may be updating their stocks of anti-tank guns firing APFSDS rounds.

MECHANISED COMBAT

speed and direction, the altitude of the tank from a sensor in the turret roof, and information on the 'bend' of the gun. For example, if the tank is operating in rain and the barrel is hot, it will distort, fractionally perhaps, but enough to affect the weapon's accuracy over long ranges. The gunner meanwhile manually sets information on ammunition type, barrel wear, barometric pressure and ammunition temperature.

Deliveries of the improved M1A1 Abrams began in 1985. The M1A1 has a number of detail improvements, but the most notable change is in its armament. The 120-mm gun is a Rheinmetall design, which has a smooth bore optimised for firing fin-stabilised discarding sabot rounds.

Browning M2 .50-cal heavy barrel machine-gun
Mounted above the tank commander's station, this has an elevation of +65° and has 360° traverse. 1,000 rounds of .50-cal ammunition are carried.

Armoured doors
These separate the crew from the shells stored in the turret bustle and should save them from the consequences of a hit on the back of the turret.

Blow-off panels
The top of the ammunition storage area has panels designed to blow off if the area is hit. This dissipates the pressure, and hopefully prevents the explosion of the tank's ammunition and the loss of the vehicle.

Turret bustle
This carries 44 105-mm rounds in the M1 or 34 120-mm rounds in the M1A1. Plastic rods and bars separate each shell to prevent the explosion of one setting off the others.

Gas turbine
There are few differences between the engine fitted to the M1 and the M1A1; it is designed primarily to use diesel or kerosene-based fuel, but can use petrol in an emergency.

M250 smoke discharger
This British-designed six-barrelled smoke discharger is fitted to either side of the turret.

Driver
With the vehicle 'buttoned up', i.e. the hatches closed, the driver is in a semi-reclining position. He steers the M1 with a motorcycle-type T-bar with a twist grip throttle at both ends.

31

MECHANISED COMBAT

Night operation

The M1 can operate effectively at night. The driver has an image intensifying periscope for night driving, and the gunner has a thermal imaging system which projects infra-red detected imagery directly into the eyepiece of his day sight. In addition, the sight displays target range information and indicates when the weapon is ready to fire.

That all this highly sophisticated (and highly expensive) electronic fire control represents a sound investment can be demonstrated by the M1's consistently exceptional performance in gunnery contests. But however lethal a tank's first-round kill probability, it will not be effective unless the vehicle can survive and prove reliable on a high-intensity battlefield. The M1 was designed to slug it out with anything the Russians might field on NATO's Central Front, and keep the technological lead into the next century. That means protection as well as firepower.

Strong armour

The M1's hull and turret are made of anti-tank missile-resistant composite armour comprising laminated layers of metal, plastics, and ceramics. Latest variants have greatly increased protection provided by ultra-dense DU (non-radioactive Depleted Uranium) armour. Design for survivability includes internal bulkheads with sliding armour doors isolating the crew from the main armament stowage. Blow-off panels are designed to channel blast outwards, while an automatic Halon fire-extinguishing system is designed to tackle internal fire instantaneously. The M1A1 also has an integrated NBC system, providing the crew with scrubbed air for breathing.

One of the controversial aspects of the M1's design during its development phase was the decision to use a gas turbine powerplant rather than the traditional diesel. In pre-production trials the vehicle encountered severe problems with transmission, track throwing, fuel supply and turbine blade failures largely due to dust ingestion.

Acceleration and speed

These problems were more or less solved by the time the tank entered operational service with the US Army, although range and track life (1,300–1,800 km [800–1,110 miles]) are still below original design parameters. The Avco-Lycoming AGT-1500 gas turbine driving through an automatic transmission delivers a power to weight ratio of 27 horsepower per tonne, affording this massive vehicle remarkable acceleration and high speed on road and cross-country.

The M1/M1A1 Abrams main battle tank has been criticised for being too sophisticated and too expensive. But after a few initial stumbles, it is standing up well to real operational conditions.

The M1 was a smashing success on its combat debut. It was the heart of the coalition armoured force that destroyed Saddam Hussein's army in Kuwait and southern Iraq, totally outclassing the Soviet-built armour of the Iraqi army. That performance came as no surprise to the US Army. On a tour of the United Arab Emirates six months previously, a four-man US Army crew took an Abrams through a firing demonstration against targets half the size of those normally used in NATO exercises. Firing from stationary positions and on the move, the M1's performance was outstanding. At ranges between 1,300 and 3,000 m (4,300 and 9,800 ft), by day and night, the Abrams scored first round hits on 38 out of 38 targets!

With 8,000 tanks in service, the M1 Abrams will form the backbone of US armoured forces well into the 21st century. Agile, powerfully armed, and superbly protected, it will remain for much of that period one of the world's most powerful tanks.

M1 Abrams form up in column of route somewhere in Germany. With its combination of power, protection, and mobility, the M1 has some claim to be the best tank in the world.

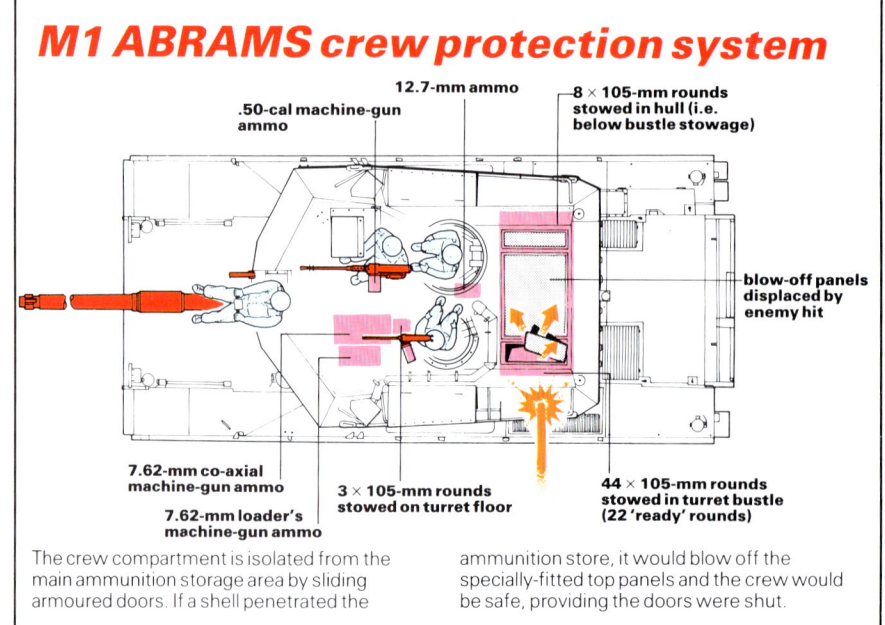

M1 ABRAMS crew protection system

The crew compartment is isolated from the main ammunition storage area by sliding armoured doors. If a shell penetrated the ammunition store, it would blow off the specially-fitted top panels and the crew would be safe, providing the doors were shut.

EYEWITNESS

World-beating M1

"You know, a lot of what was said about the M1 before the war, that it was too complicated and was out of action more than it was in service, was a load of bull. Serviceability was greater than 90 per cent. And all those Iraqi tanks that were going to hammer at us from behind their awesome field fortifications.... Well, we found we could kill T-62s and T-72s even when they hid behind sand berms by firing sabots right through the berms and into the tanks behind. Plus, our guns could outdistance theirs by 1,000 m [3,280 ft]. I know of a couple of M1s that were hit by rounds from T-72s, and sustained no damage. That might explain why only four out of 2,000 M1A1s were disabled during the war, and only four more suffered less serious battle damage. You look at that kind of performance from the best tank in the world and I guess that it's no surprise that the ground war was over in a hurry."

**Lieutenant Mike Drury
Platoon leader, 1st Armored Division
Saudi Arabia/Iraq/Kuwait**

MECHANISED COMBAT

MOBILE DEFENCE

A squadron of M1 Abrams tanks prepares to move out to the battlegroup concentration area in preparation for a mobile defence exercise in the area of the Fulda Gap.

When the main threat to your position comes from numerically superior enemy armoured forces, a successful defence will depend more than anything else on how you deploy, co-ordinate and control the anti-armour weapons in your battlegroup. Use anti-tank guided weapons (ATGW) wherever the ground, their rate of fire and their minimum ranges allow them to be the mainstay of your defence against armoured personnel carriers. You can use tanks, if necessary, to cover any gaps or to reinforce the defences on the most dangerous approaches.

Don't tie tanks to purely defensive tasks. By using ATGW and Warrior mechanised infantry combat vehicles (MICVs) properly, you should be able to release most, if not all, your tanks to operate offensively. You can see from this approach how positional and mobile defence are inextricably

PRINCIPLES OF MOBILE DEFENCE

1. Offensive action
You cannot win battles unless you attack. Having identified the main axes of enemy thrust, block and destroy him.

2. Command and control
You must have the means to control a highly fluid mechanised battle where accurate information can be passed back fast enough for the right command decisions to be made.

3. Maintaining a reserve
Hoard as much armour, artillery and engineer assets as possible for your mobile reserve. All local commanders fighting the positional defensive battle should also keep a reserve.

4. Concentration of force
Concentrate your mobile reserve in superior numbers.

5. Firm base
The offensive action by the mobile reserve must be launched from a firm base.

6. Mobility
Concentrate the mobile reserve quickly at any point the enemy armour breaks through.

MECHANISED COMBAT

linked. Purely positional defence is unlikely to bring you victory. Purely mobile defence is unlikely to succeed without the anchor or pivot of a firmly held 'shoulder'.

Steal the initiative

Mobile defence uses a combination of offensive, defensive and delaying actions to defeat an enemy attack. The art of successful mobile defence lies in using relatively small forces forward and manoeuvring, with support from fire and obstacles, to wrest the initiative from the attacker.

This means that you will need mobility equal to, or even greater than, that of the enemy. You must, above all, be able to form a reserve that is strong enough to launch a decisive counter-attack. Almost inevitably, you will only be able to do this by thinning out forces committed elsewhere. You can't carry out a successful mobile defence unless you are prepared to accept gaps or to lose some terrain.

Mutual support is vital in defence at platoon and company level. Ideally, you should also try to achieve mutual support within battlegroups. However, at brigade and divisional level,

The gunner of an M60 tank selects a 105-mm main armament round for loading. There is now some doubt as to how well the M60 will perform when pitted against the new generation of Russian tanks.

and often at battlegroup level, the need to form a large reserve means that you will have to accept gaps to maintain depth and your reserves. Mutual support is clearly not possible in these circumstances.

On the move

While you'll always need to hold ground to form a framework for your defence, mobile defence is fought through manoeuvres. You have to occupy alternative fire positions and counter-penetration positions in depth, and counter-attack. By keeping reserves in depth you will have the time and space to identify enemy breakthrough, and choose the routes that put you in a position to counter-penetrate or counter-attack. You will then have to calculate how long it will take to get you to your new position, whether to move by day or night and what forces you'll need to complete the task when you get there.

HOW TO SITE ANTI-TANK GUIDED WEAPONS

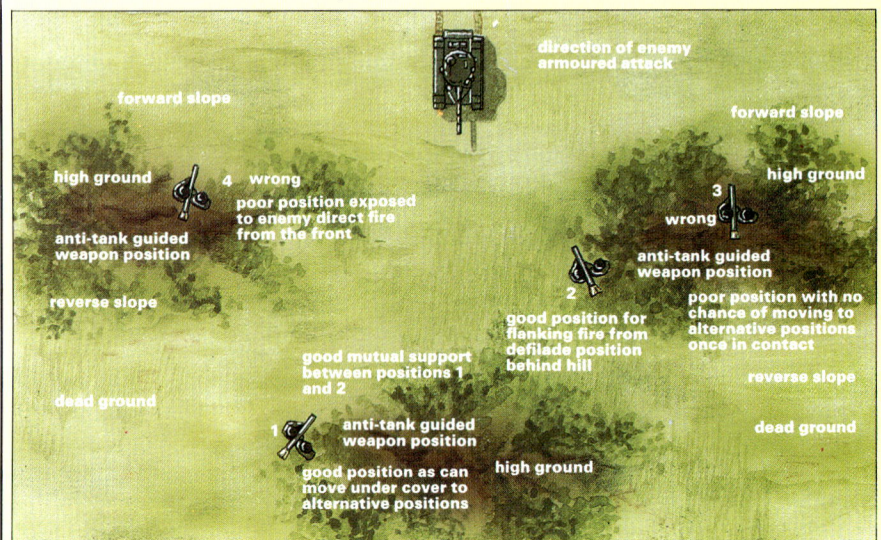

Position anti-tank guided weapons so that they have maximum cover and concealment from aerial and ground observation, good fields of fire out to their maximum range and mutual support from other weapons. Avoid conspicuous terrain features. Lastly, and most importantly, site them to engage the enemy from a flank.

Alternative positions

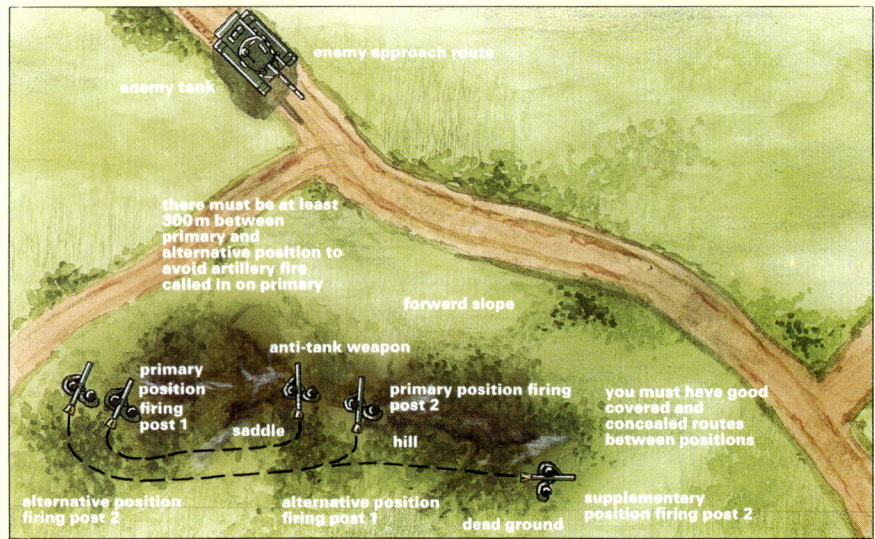

The key to the mobile battle is the strength of the position defence on which it is based. This in turn is largely dependent on the success of the anti-armour guided weapons battle. Recce and dig in several firing positions for each firing post: you will not survive long if you continue to engage targets from one firing point.

MECHANISED COMBAT

Whenever possible, concentrate your armoured reserve and use it to attack. If necessary, carry out different tasks in succession rather than dissipating your limited resources in wasted 'penny packets'. Once you have decided to commit your reserve, speed of reaction is vital. In counter-attack or counter-penetration, minutes or even seconds may decide the issue. And, once he has committed his reserve, the battle-group commander must immediately try to create another reserve, however small.

Your aim in mobile or manoeuvre warfare must be to outwit, outmanoeuvre and ultimately destroy the enemy, either by enticing him into a trap or by surprising him with a counter-move – or both. It is not good enough just to withstand his attack.

At your service

Every battlegroup is equipped with the Challenger tank and the Warrior MICV. Both are designed for mobile warfare. Warrior has the power to keep up with Challenger across country and has a top road speed of 75 km/h (47 mph); it is armed with a 30-mm RARDEN cannon and a 7.62-mm Hughes Chain Gun. It will give you effective fire support both in defence and in the assault.

Combat engineers are also armed and equipped to support mobile warfare. The Armoured Vehicle Royal Engineers (AVRE) has a 'dozer blade to fill in anti-tank ditches and carries a fascine to bridge larger gaps. The Giant Viper is designed to create channels through minefields by laying an explosive hose across the minefield and then detonating it. Engineer support is as important for your mobility as it is to counter enemy movement.

The attack helicopter is a key weapon. The only real problem is their vulnerability: fire-and-forget missiles such as Hellfire launched from an Apache seen here are a great step forward.

You must be able to move your forces freely round the battlefield at speed to counter enemy penetration. The success of the battle will therefore depend on who best uses his engineer assets. The Chieftain armoured bridgelayer can span a 22-m (72-ft) gap.

The covering force, once it has withdrawn from the covering force battle, will be used to bolster the mobile reserves and as such will depend on vehicles such as this Stalwart for resupply in forward areas.

Left: TOW can destroy armour out to 3,750 m (12,300 ft), but the operator has to track the target for the whole flight time and, although there is a 95 per cent hit probability in ideal conditions, enemy suppressive artillery fire could considerably reduce this.

Above: An A-4 Skyhawk flies past a huge secondary explosion produced by the lead aircraft's bomb run on a simulated fuel dump target. It is vital to maintain air superiority over the area you select for your counter-attack.

MECHANISED COMBAT

Perhaps the most important tool of mobile warfare is the helicopter. The attack helicopter gives commanders the means to counter armoured thrusts by launching devastating ATGW attacks with large numbers of helicopters. The British Lynx, for instance, carries eight TOW missiles in eight launchers and a reload of a further eight missiles in the aircraft. The missile, which has a range of 3,750 m (12,300 ft), can kill all known types of Russian-built tanks. Given surprise and good tactics, just a squadron of Lynx helicopters can stop an enemy tank regiment. The US Apache and Soviet-built Hind helicopters are equally, if not more, effective. Support helicopters such as the Chinook and the Puma, both in service with the RAF, can lift air-mobile troops and drop them in position to block an armoured breakthrough.

But air-mobile troops are only strong enough to block a breakthrough; it takes an armoured formation to destroy one. So mobile warfare is all about getting the right combination of troops to the right place in time. It also means accepting that NATO does not have enough forces to occupy positions in depth all along the NATO front line. Gaps will have to be left; penetration by the enemy will have to be accepted.

Because he has the initiative, the enemy can concentrate his forces where he wants. This means he can achieve a local superiority of perhaps eight or 10 to one. There will be nothing you can do about that but accept penetration. But that does not mean accepting defeat or even a breakthrough.

If you keep sufficient reserves in depth and make sure that they are sufficiently mobile – on tracks, in the air, or a combination of both – you will be able to block the enemy or counter-attack him or, best of all, mount what NATO now calls a 'counter-stroke' against him.

Combination of obstacles

This means enticing him into an area that you have prepared for him. Use a combination of natural and artificial obstacles to channel him into a position where you can launch an attack against his flank or rear. Mobile warfare thus takes advantage of the enemy's desire to press on regardless. Inevitably, he will leave his flanks open, and this is where you can hit him. If you can cause sufficient destruction and dislocation, you will bring his attack to a halt, cut his lines of communication and leave his armoured spearhead stranded like a great whale.

The British Army now has the tactics and the equipment to fight a mobile war. By putting the offence back into defence you stand a chance of winning!

Above: An American-made M60 Main Battle Tank of the Israeli army fitted with reactive armour to protect it against the hollow charge warheads of infantry anti-tank rockets.

Below: An M1 Abrams fires its 105-mm main armament. This has proved perfectly capable of knocking out Soviet-built T-72s, but most US Army M1s are now armed with 120-mm weapons.

MECHANISED COMBAT

STORM THE LINE WITH CHALLENGER

Below: A Challenger storms across the desert, its powerful 120-mm gun towards the enemy. The desert is ideal combat terrain for tanks, allowing them to use their mobility and long range.

Bottom: Developed from the 1950s-technology Chieftain tank, the Challenger has been the subject of a lot of criticism. It silenced most critics with an excellent performance in the Gulf.

Challenger tanks formed the steel core of the British armoured division that fought in the Gulf. Armed with a 120-mm rifled gun capable of penetrating 400 mm (15.75 in) of armour plate, Challenger can destroy any Russian-built tank with a single hit.

Challenger is protected by Chobham armour – blocks of dense ceramic material held in a resin matrix sandwiched between two layers of steel plate. This incredibly expensive material is equivalent to 500 mm (19.75 in) of conventional armour against high velocity armour-piercing shells. Against the HEAT warheads of infantry anti-tank weapons it is equivalent to 800 mm (31.5 in), making it all but immune to hand-held rockets and wire-guided missiles.

Challenger's tremendous firepower and formidable protection make it one of the best tanks in service today. But it was not designed for the British Army. The 'Desert Rats' only went to war with Challenger because thousands of Iranians rioted in Tehran and overthrew the Shah.

Iranian cancellation

Challenger was developed for the Imperial Iranian army. After Ayatollah Khomeini seized power the Shah's order was cancelled and the Royal Ordnance factory in Leeds was left without work. The British government ordered 243 Challengers, saving the jobs of a highly skilled workforce and giving the British Army a superb tank 10 years before any replacement was scheduled.

Challenger is powered by a Rolls-Royce Condor 12V-1200 turbocharged diesel engine. Despite its heavy combat weight of 60 tonnes (59 tons), Challenger has a respectable power-to-weight ratio and can sustain a maximum road speed of 56 km/h (35 mph).

The 120-mm L11A5 rifled gun fires a variety of ammunition. The primary tank killing round is the fin-stabilised armour-piercing projectile. High Explosive (HE) and HESH (High Explo-

37

MECHANISED COMBAT

Challenger is one of the world's most heavily protected tanks, and its 'Chobham' laminated steel/plastic/ceramic armour is designed to defeat enemy shaped-charge warheads.

Armour-Piercing, Fin-Stabilized, Discarding Sabot (APFSDS)

The most effective anti-tank round in the Challenger's armoury is the APFSDS. This works by projecting a small calibre penetrating dart at very high speed. It is surrounded by a lightly fitting sleeve known as a sabot, which fills the bore of the gun as it is fired. Once free of the barrel, the sabot falls away leaving the penetrator to fly free to the target. The high-density penetrator is stabilised in flight by fins, and has so much kinetic energy that it smashes through the tank's steel hull. The immense shock will detonate secondary explosions in the target's fuel and ammunition stores.

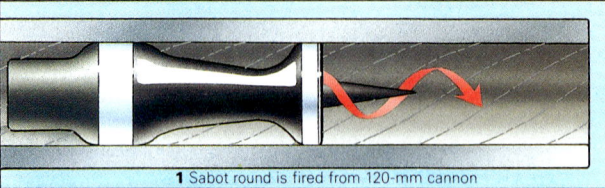

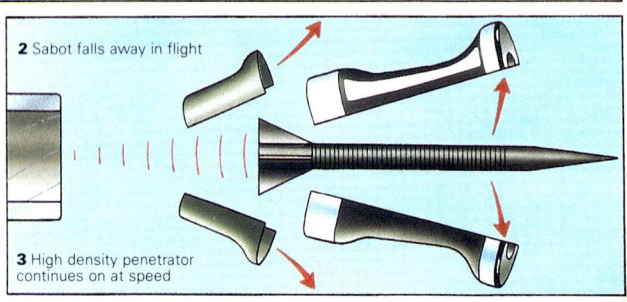

1. Sabot round is fired from 120-mm cannon
2. Sabot falls away in flight
3. High density penetrator continues on at speed
4. Penetrator smashes through armour of target tank
5. Shock causes secondary explosions, destroying target

sive Squash Head) are also carried.

A HESH round contains an explosive charge that pancakes onto the target's armour before detonating. Shock waves caused by the resulting explosion cause the inner surface of the target to fracture, flake off and fly around until contact is made with a crew member or, more drastically, stored ammunition.

The more advanced APDS and APFSDS rounds consist of a sub-calibre projectile with a sabot, or lightly sectioned 'sleeve', fitting the residue of the bore. Once the round is fired the sabot splits and falls away, leaving the projectile to travel at very high speed until it strikes and forces its way through the target.

An L7A2 7.62-mm machine-gun is fitted coaxially with the main armament and another 7.62-mm GPMG is mounted on the commander's cupola, aimed and fired from within the turret. Recent reports suggest that this

Rolls-Royce Condor 12V 1200 engine
Based on proven conventional components, this is an excellent engine with low specific fuel consumption. It uses high efficiency turbochargers and can develop 1200bhp at 2300 revolutions per minute. The same engine is fitted to the Khalid tank, essentially a Chieftain with a Challenger powerpack, sold to Jordan.

120-mm L11A5 rifled gun
Challenger's main armament is fitted with a thermal sleeve, fume extractor and muzzle reference system. When it enters service, the L30 high-pressure tank gun will be retro-fitted to the Challenger force.

Single pin track
Challenger's track is not interchangeable with Chieftain's. It is a single-pin track with removable rubber pads, although this may be replaced with a new track which offers less rolling resistance and has a longer life expectancy.

Left: The high-density penetrator has immense kinetic energy, which enables it to smash its way through opposing armour, wreaking havoc within the enemy vehicle.

MECHANISED COMBAT

may be replaced by an old pre-war design Vickers 12.7-mm machine-gun. On the face of it the idea of fitting a tank of the 1990s with a gun of the 1930s may seem ridiculous, but this larger weapon would be far more effective against Russian helicopters.

Secret armour

Chobham armour is as secret as it is revolutionary. Even now, some five years after Challenger made its first appearance with NATO on Exercise Lionheart, little if any open source material exists on the subject.

Conventional spaced or laminated armour can defeat the HESH round, but it is of limited defence against the newer high-velocity APDS rounds, and it is to defeat these that Chobham armour was developed.

Inside the Challenger

While Challenger has a less powerful engine than its American and German contemporaries, Challenger's superb transmission and hydro-pneumatic suspension allow it to travel just as fast across country as its American and German contemporaries.

Gunner
Challengers are fitted with TOGS (Thermal Observation and Gunnery System) which lives in an armoured barbette on the right of the turret and provides separate outputs for commander and gunner.

Commander
He has a modified No. 15 cupola which is fitted either with a day sight or image intensifier for night combat. Nine periscopes provide the commander with all-round vision.

Commander's machine-gun
The commander's cupola has a 7.62-mm L37A2 machine-gun for rather optimistic anti-aircraft defence. The co-axial 7.62-mm L8A2 machine-gun is of much greater value since it can be fired from within the vehicle.

Loader
His task is to load the Challenger's 64 rounds of 120-mm ammunition: generally 20 'fin' and 44 HESH and smoke rounds. The ammunition is separate. Each charge storage position contains either one charge for an APDSFS round or two for HESH/smoke.

Chobham armour
Still highly classified, Chobham armour is named after the MoD establishment where it was first developed. It is particularly effective against chemical energy attack, so infantry anti-tank rockets and anti-tank guided missiles present much less of a threat to Challenger than they do to Chieftain.

Driver
The driver can swing his single-piece hatch forward so he can drive with his head out. He has a wide angle periscope for day driving, which can be replaced by a Pilkington passive night-sight for driving in conditions of darkness. In an emergency the driver can escape through the fighting compartment.

MECHANISED COMBAT

The structure is known to consist of numerous layers of metals, ceramics and plastics designed to absorb and break up the impact of the high-speed core of the APDS round. First sight of Challenger will show not only how immensely thick are the slabs of armour, especially on the forward chassis and turret, but also how angled they are. Indeed, it is suggested that the 60 per cent angle of the turret armour, not found on earlier British tanks, more than doubles crew protection against all conventional anti-tank weapons fired other than at extremely close range.

Extra weight

The exact thickness of the armour can only be assumed, but it is interesting to note that Challenger is some five tonnes heavier than Chieftain Mk V. Although part of this weight difference is the result of the larger engine, Chobham armour is as heavy as it is chunky.

Thermal imaging sights and a comprehensive NBC system are fitted as standard, and a bolt-on hydro-pneumatic suspension system greatly facilitates maintenance and even replacement in the field. One of the most versatile of tanks in service anywhere, Challenger can climb gradients of 60 per cent, overcome vertical obstacles as high as 0.9 m (3 ft) and cross trenches up to 3 m (10 ft) wide.

Above: British tank crewmen camouflage their Challenger with netting in a laager somewhere in the Saudi desert shortly before the start of the massive Coalition ground offensive into Kuwait and Iraq. Once on the move, the Challengers formed a fast-moving spearhead through Iraq to cut off and engage the Iraqi Republican Guard.

Above: Challengers head off into the deserts of southern Iraq as part of the massive armoured offensive that was to rip apart Iraq's Republican Guard in less than 100 hours.

Right: The problem with serving in tanks in the intense desert sun is that they get very hot inside. As a result, when crewmen get a chance to rest in the shade, they take it!

EYEWITNESS

Challenger triumphant

"By the time we'd spent a month or two in the desert, we were dead keen to get into action. Once we got going, it was a relief. A lot of fighting took place at night. Our TOGS, or thermal observation gunnery sight, gave us a real advantage since we could engage the enemy long before they could even see us. I heard that one Challenger took out an Iraqi tank at a range of 5,000 m (16,400 ft), which is way beyond normal fighting range. One problem with fighting at long range was that we'd slam shell after shell into a tank, before realising that the Iraqis had baled out long before and we were engaging an empty vehicle!

"The Iraqi tanks weren't too advanced technically, so we didn't need to use high-density armour-piercing rounds. Our rifled gun gave us the option to fire HESH — a high-explosive squash head — which had a slightly longer range than a fin round. That was something the Americans with their smoothbore 120s could not do. HESH was easily capable of taking out a T-62. It was a good bunker buster as well."

Trooper McAuslan, 7th Armoured Brigade, Saudi Arabia

MECHANISED COMBAT

WORKING WITH TANKS

In most combat situations, tanks and infantry are indispensable to each other: each can fulfil certain tasks that the other cannot. Tanks are more suited to operating in open countryside, where they can engage targets at long range; but infantry prefer urban or wooded areas, where they are less vulnerable and where their shorter-range weapons are more effective.

Tanks can cover infantry while they are in the open, and infantry can protect tanks while they are at their most vulnerable from close-range ambush in wooded or built-up areas. So you and your supporting tanks must know how to work closely together.

Sometimes you will want to direct each other's fire onto enemy targets. Usually it will be you, the infantry commander, who will want to direct tank fire onto targets that are obstructing your advance or causing casualties to your men. You may, for instance, be finding it hard to neutralise a well dug-in machine-gun. Tanks will often spot such targets before you do, and engage them without your guidance. You, of course, can do the same for the

TASKS FOR TANKS AND INFANTRY

These are the tasks an armour heavy battlegroup would be expected to take on. 'Armour heavy' simply means that there are two squadrons of tanks and one company of infantry.

1. Rapid advance to contact with the enemy, or following up a deliberate withdrawal.
2. Assault and destruction of enemy defended positions.
3. Exploitation of weaknesses in the enemy front line, penetration and pursuit.
4. Counter-attack and counter-penetration.
5. Aggressive reconnaissance by day and night.
6. Diversionary operations and flank protection.
7. Screening or covering force operations protecting the main defensive position and/or buying time for its completion.

An M67 flamethrowing tank of the US Marine Corps suppresses an area suspected of containing a VC sniper. A tank's massive firepower is the result of the combination of fire from main armament and machine-guns. Correct use of tank support once in contact will minimise your infantry casualties. In return, you must protect the tanks from enemy anti-tank weapons.

Tanks are great for cracking hard pinpoint targets like bunkers, other tanks and APCs. These US Marines are using an M48 to engage a well dug-in NVA position. American use of armour in close country in Vietnam was surprisingly effective.

41

MECHANISED COMBAT

Austrian infantry charges past a knocked-out tank on exercise. Armour is extremely vulnerable in close country without adequate infantry protection. In war, keep away from tank hulks: the ammo may cook off, and they tend to draw fire.

tanks by dealing with anti-tank weapons. But if a target has not been spotted by you or your supporting tanks, you must have a method of indicating targets to each other.

Quick transmission

Usually you will need to transmit your message quickly, so you cannot use the slow and complicated Target Grid Procedure. Similarly, you cannot use smoke since there is likely to be a lot about already, and anyway it might obscure the target.

So how do you talk to a tank commander? Basically, you have three choices: by radio, by tank telephone or by personal contact. Each method has advantages in different circumstances. The best is personal contact but, as this usually requires you to climb onto the tank, it is not always

ARMOUR AND INFANTRY COOPERATION

The roles of armour and infantry although different, are complementary. Tanks are suitable for some tasks, and infantry for others; in order to get the best results each must know the other's capabilities and limitations. Armour and infantry usually work closely together, and the ratio of tanks and infantry is varied according to the task in hand. For example, in attack you need more tanks; in defence, you need more infantry.

Holding ground
Tanks cannot hold ground. They may help in evicting the enemy, but it is the infantry who keep the enemy off a particular piece of real estate.

Personal contact
This is the best method for fast and efficient fire control, but is risky as the tank will draw enemy small-arms fire and when buttoned up it is virtually blind within 15 m (49 ft), so there is a chance you will get run over.

Shock action
The aggressive use of tanks exploiting all their tactical characteristics of firepower, mobility, flexibility and armoured protection produces 'shock action', and a devastating effect on enemy morale.

Logistic support
One of the drawbacks of the tank is that it cannot go without logistic support for days, like the infantry. After an average day's battle the chieftain will require 20 jerry cans of fuel and 40 rounds of ammo and considerable servicing time.

MECHANISED COMBAT

advisable. So usually you have to use the radio or the tank telephone.

Both the Chieftain and Challenger Main Battle Tanks have a small box attached to the rear with a telephone that connects you to the tank commander. Being situated at the back of the tank, you are automatically covered, but make sure that the driver is not about to put the tank into reverse!

Call sign

If you have to use the radio you will need to know the call sign of the tank you want to talk to: this will be painted on its turret or hull.

Within each squadron there are four troops of tanks, and within each troop there are four tanks. 'A' Squadron's call sign is 1, 'B' Squadron's 2 and so on. So the 2nd Troop leader in 'A'

This is tank country. Here the infantry is very vulnerable without the tank support. Tanks and infantry usually fight together in mixed units made up of differing combinations called company squadron groups.

Battalions in Germany no longer fight as separate units but as the 'all arms' battlegroup made up of tanks, infantry, engineers, signals, REME support and dedicated artillery. This is a company squadron group of Chieftains and APCs.

Night vision devices and optics
Modern tanks have excellent night fighting devices and optics which can be used to identify infantry targets. Do not forget the searchlight, which may also be of use in the right tactical situation.

Surprise
The size, weight and noise of the main battle tank may make surprise difficult to achieve. However, good forward planning, for example covering the noise of movement with artillery and smoke, may solve the problem.

Sensitivity to ground
Unlike infantry, tanks have problems crossing boggy or rocky ground, and steep slopes, thick mature woodland, rivers and minefields are effective barriers.

Tank telephone
Most tanks are fitted with a tank telephone on the outside rear of the tank, but most crews have ripped them out to discourage the infantry using them (there have been some nasty accidents on exercise).

Radio
This is the usual method of communication, and the safest as you can keep your distance.

MECHANISED COMBAT

Indicating targets to tanks

When you are working with tanks rather than artillery you have to be in contact with the individual tank commander who controls the gun fire. Once the tank is buttoned up it has limited vision, so you must be its eyes and ears: spotting targets, giving target indications and advising the type and duration of fire.

There are several methods of indication:

1 Reference point method
This works in the same way as giving fire control orders to a section. The platoon or section commander spots a target and requests help from a supporting tank on the radio. The tank is operating on the same radio frequency as the infantry radio net and is called using the individual tank's callsign, which in this case is Tango One Three. The platoon commander's callsign is India One One Alpha.

Inf: "Hullo T13, this is I11A, target, over"

Tank: "T13, send, over". The tank commander is ready to receive the message, in this case the target information.

Inf: "I11A, lone tree [a reference point briefed to all at the end of the last tactical bound], **go right four o'clock** [the infantryman is using the clock ray method to indicate to the tank commander where the enemy is], **400/1,310** [range in m/ft to the actual target], **machine-gun in hedgerow** (you must tell the tank commander what he is looking for so he can identify it and select the right type of ammo], **destroy** [this is what you want done to the target], **over**."

Tank: "T13, wilco, out".
"Wilco" stands for "will comply", which means the order is understood and will be actioned. Watch out for fall of shot as he is about to plaster the target with machine-gun fire and a splash of HE to complete the job.

2 Gun barrel method
To use this method you tell the gunner where the tank gun is pointing in relation to the target.

Inf: "Hullo T13, this is I11A, target, over."

Tank: "T13, send, over."

Inf: "11A, gun barrel [method of indication of target], **quarter right** [move the gun barrel right through 45 degrees], **800/2,620** [the m/ft range to the target], **small copse, identify, over.**"
The target is well camouflaged and difficult to see, so the infantry commander gives the indication in two halves, enabling the tank to use its excellent optics to positively identify the target.

Tank: "T13, identified, over."
The tank commander swings the turret quarter right and looks out over the gun barrel at 800 m (2,620 ft) for the copse. If he cannot see the copse then he reports "Not seen", and the infantry sends another indication.

Inf: "I11A from copse, right, four o'clock, enemy bunker, neutralise and cease fire on my order, over."

Tank: "T13, wilco, out."

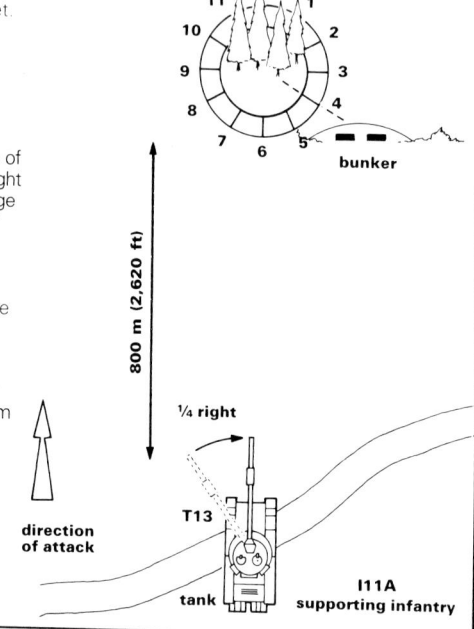

3 Burst for reference fired by the tank
Where a target is particularly difficult to indicate, ask the tank to fire a burst from its maching-gun in the general direction of the target and then correct by watching where the burst goes.

Inf: "Hullo T13, this is I11A, target, over."

Tank: "T13, send, over."

Inf: "I11A, gun barrel, quarter left, 600 [1,970 ft], fire burst for reference, over."

Tank: "T13, wilco, wait out."
You will have to wait while he identifies the target.

Tank: "I11A, this is T13, shot out."
This warns the infantry to look out for the burst he is about to fire to note where it lands.

Inf: "Hullo T13, this is I11A, from last burst right 100 (328 ft) drop 50 (164 ft), enemy in bunker, neutralise, I am attacking left flank, over."
This corrects the tank machine-gun fire onto the target and warns the tank to expect them to attack the position from the left. When the infantry are almost on the enemy, the infantry would call for rapid fire as the assault goes in and would then tell the tank to switch fire to targets in depth when the infantry have reached the bunker.

Tank: "T13, wilco, out."
The tank will fire at a steady rate, keeping the enemy's heads down and conserving ammo as the infantry approach their FUP for the attack. Then, as they assault, the tank will fire longer bursts to suppress the bunker as the infantry move to grenade or satchel charge posting distance.

Swedish troops debus to assault an objective supported by fire from their APCs and S tanks. The tanks shoot them onto the objective and then switch fire to any positions in depth.

Squadron will be 12. His Troop Sergeant's tank will be 12A, and the other two tanks commanded by Corporals will be 12B and 12C. So immediately you look at the tanks you can tell which tank he is in and which troop and squadron he belongs to.

Once you have attracted the tank commander's attention, direct his attention to the target. You have three ways of doing this.

Reference points
First, you can use reference points. If possible, arrange these before an operation. You must select them by looking at the battleground and not from a map. Choose clearly defined features that will leave no room for confusion.

You can use your chosen reference points to guide the tank commander's eyes from it to the target via a succession of landmarks using the clockface method.

Tank gun barrel method
If you cannot use the reference point method, the tank gun barrel can be used as a datum line instead. You will normally be able to see the direction of the gun barrel, and so you can guide it 'quarter RIGHT' or 'half LEFT' onto the target. This is a simple but effective method that requires no planning before an operation.

Shell strike method
If it is impossible to use either of these methods, you can use the burst of a shell from a tank, or the strike and tracer from an infantry or tank machine-gun, as a datum point from which corrections can be given. If you are going to use one of your machine-guns to mark a target, warn the tank commander when you are going to fire and also indicate the direction he must look in. He must do the same for you, if he is using his main armament or a machine-gun to indicate a target to you.

MECHANISED COMBAT

Blast them with the MLRS

Above: A Royal Artillery MLRS engages Iraqi positions during the Allied ground offensive. The Iraqis called MLRS 'steel rain': they had no answer to a hail of deadly accurate rockets each containing 688 grenade-like bomblets.

Below: A US Army MLRS in firing position. The moment Iraqi heavy guns opened fire, their positions were computed by Firefinder radar. MLRS batteries began to fire on the enemy artillery before the Iraqis' shells had even landed!

As British ground troops attacked enemy positions in southern Iraq, the Royal Artillery pulverised Iraqi trenches with its latest weapon. British gunners are trading in their heavy guns for MLRS (Multiple Launch Rocket Systems). The internationally-developed MLRS can deliver a devastating concentration of fire. Over a short period a single launcher can deluge a target with more high explosive than several batteries of conventional artillery. And in high-intensity operations, that ability is exceptionally important.

Multiple rocket-launchers are ideal offensive weapons systems, able to shatter enemy defences immediately prior to an attack. They are ideal for the delivery of intense concentrations of smoke to conceal your advance, or chemical ammunition such as blood-agents which are far more lethal if delivered all in one go rather than steadily built up by a conventional artillery barrage.

Mounted on a tracked self-propelled loader-launcher vehicle (SPLL)

45

MECHANISED COMBAT

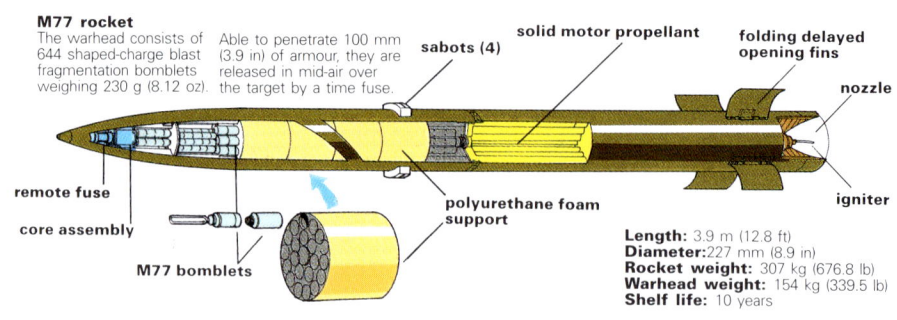

M77 rocket
The warhead consists of 644 shaped-charge blast fragmentation bomblets weighing 230 g (8.12 oz). Able to penetrate 100 mm (3.9 in) of armour, they are released in mid-air over the target by a time fuse.

Labels: remote fuse, core assembly, M77 bomblets, sabots (4), polyurethane foam support, solid motor propellant, folding delayed opening fins, nozzle, igniter

Length: 3.9 m (12.8 ft)
Diameter: 227 mm (8.9 in)
Rocket weight: 307 kg (676.8 lb)
Warhead weight: 154 kg (339.5 lb)
Shelf life: 10 years

Left: If the MLRS launcher is loaded it is possible for one man to run the engagement sequence on his own. All 12 rockets can be ripple-fired in under a minute.

Above: Inside the MLRS's DPICM (Dual-Purpose Improved Conventional Munitions). The explosion of 688 bomblets proved highly effective against Iraqi guns.

Above and right: MLRS is loaded from pre-packed pallets of six rockets. The operation is largely automated and can be accomplished in only a few minutes. Ammunition can be pre-dumped near likely firing sites.

based on the M2 Bradley chassis, MLRS has the mobility and speed to keep up with armoured units. Its ability to halt, fire 12 independently-aimed missiles and withdraw, all in the space of 90 seconds, makes it a formidable weapon.

The crew of three (commander, driver and gunner) can reload with two rocket containers, each with six pre-loaded tubes, without leaving the protection of their armoured cab. Each vehicle is equipped with its own inertial navigation system and computer which can be independently programmed by a front-line forward observation officer direct, thus saving considerable time. The crew can ripple-fire from two to 12 rounds in less than one minute, the fire control automatically re-aiming after each shot.

MLRS is designed to fire any of three types of submunition: bomblets, anti-armour minelets and guided sub-missiles. There is a suggestion that a binary chemical warhead may be under development, but at the moment this is a matter of pure conjecture.

The M-77 bomblet, now in service with the Royal Artillery, is of greatest use against unprotected soft-skinned vehicles, armoured personnel carriers and towed artillery. Six hundred and eighty-eight bomblets are packed into each 159-kg (350-lb) missile, so each MLRS is capable of simultaneously discharging 8,256 bomblets over an area. To give some indication of the potential havoc this could cause, it has been said that a battery of launchers could obliterate an area the size of four football pitches and that only two launchers would be required to destroy an enemy gun battery dug in up to 30 km (18.75 miles) away.

Although the bomblet has obvious advantages when employed against

MECHANISED COMBAT

Inside the MLRS

In a future European war the main role of MLRS would be rapidly to lay minefields ahead of advancing enemy armour. Using the AT-2 anti-tank mine, MLRS can sow 336 mines over an area 1000 x 400 m (3,280 x 1,310 ft) in one minute.

Cab
The danger from rocket fumes is a serious problem for rocket-launcher crews. MLRS has an overpressure system in the cab to keep the fumes out, as well as a full NBC system.

Gunner
MLRS can operate with only two crew, and it is theoretically possible for one man to load and fire the system alone.

Commander

Fire control display

Fire control
The fire control system has to relay the MLRS after it has fired a rocket, since the massive blast moves the vehicle's position.

Driver

Aluminium armour
This protects the crew from small-arms fire and shell splinters, but cannot save the vehicle from a direct hit by shells or cannon.

Elevation actuators

Rockets
MLRS fires three basic types of rocket: one dispensing anti-personnel mines, one dropping anti-tank mines, and one chemical warhead (used only by the US Army). In 1985 the US Army announced that it is planning a fourth warhead called SADARM. This will carry 6 anti-tank missiles which will independently seek out their targets and fire a hypersonic velocity penetrator slug through the tank's top armour.

Boom extension actuators

Blow-off covers
The rockets are hermetically sealed in their tubes and have a shelf life of 10 years.

Launch container

Launch pod
The rockets are contained in fibreglass tubes within an aluminium six-pack weighing 2.27 tonnes (2.23 tons). MLRS batteries are supported by HEMT 8-wheel-drive trucks, each carrying four launch pods.

MECHANISED COMBAT

lightly-protected targets, it would be of little use against massed armour, and for these targets the AT-2 is being developed. Each missile consists of seven independent warheads or submunitions.

Around 1 km (0.62 miles) from the general target the main warhead disintegrates, releasing the warheads, which locate their own targets, falling with the aid of a parachute onto the victim's lightly protected roof. When this concept reaches the production stage and deployment it will clearly revolutionise the armoured battle of the future.

No matter how accurate a weapon – and the MLRS is among the most accurate in production – it will only hit its target if well aimed. Targets 30 km (18.75 miles) away are well beyond line of sight, and alternative methods of target location must be found. Traditionally spotter aircraft or helicopters were used in this role, but modern anti-aircraft weapons make this impractical. Most countries have turned to the RPV (remotely piloted vehicle) to provide the solution. Such 'aircraft', or, more correctly, drones, are either pre-programmed or controlled by radio impulses generated from the safety of the user's front line. 'Real time' signals are transmitted back to the user for translation into fire missions.

Production of the MLRS has now become an international concern. Representatives of five countries – LTVB Aerospace and Defence of the United States, Aérospatiale of France, Diehl of West Germany, Hunting Engineering of Great Britain and SNIA-BPD of Italy – are now marketing the system worldwide and already have 30 interested parties as varied as Greece,

A Royal Artillery MLRS launcher on exercise in Otterburn. The British Army is replacing heavy artillery batteries with rocket launchers.

Saudi Arabia and Pakistan. The United States has already fielded 15 MLRS batteries within the USA, in Germany and in Korea, and has reserves of over 100,000 rockets. The first European systems entered service in 1989.

The MLRS has the potential to revolutionise land warfare over the next decade and is one of the most important pieces of equipment to enter service since the end of World War II.

Left: *The cab pivots forward to allow access to the powerplant. When training to fight in Europe, the individual launchers deployed up to 5 km (3.12 miles) apart and moved immediately after firing to avoid enemy retaliation.*

Below: *The British MLRS launchers fire on Iraqi defences. The huge smoke trails instantly give away their positions.*

MECHANISED COMBAT

AMBUSHING TANKS

Afghan guerrillas pose for the camera on the wreck of a Russian BMP that they ambushed. Hit by an RPG-7 anti-tank rocket, the BMP brewed up, taking most of its crew with it. Visibility from most armoured vehicles is poor, and in forested or built-up areas the infantryman can successfully ambush the most powerful Main Battle Tank.

It is easy to think of anti-tank helicopters or other tanks as the main anti-tank weapon systems on the battlefield. This is only half right. The other tank killer is you: the infantryman, armed with a variety of portable anti-tank weapon systems. Even though some of these systems, particularly MILAN, allow you to stand off from your target and engage at ranges of up to 2,000 m (6,560 ft), most of the systems require you to be much closer. And one way of getting close to your target is to ambush it.

Normally a force of platoon strength will be given the task of carrying out a tank ambush. Before you embark on your patrol to the ambush site you must prepare your operation carefully and precisely. Make sure that you have all the information that you will need to ensure a successful ambush, especially a genuine knowledge of enemy tank tactics, capabilities and techniques.

The Russian BMP's armour is thin: this is the exit hole left by the copper slug in the RPG-7 round that penetrated the side armour, passed through the interior and out the other side.

Choose the most appropriate method of attack for the planned target and the terrain. In Germany you are most likely to carry out a tank ambush in wooded or close country or in or from a built-up area. Germany is becoming increasingly urbanised and is already heavily wooded; enemy tanks are therefore naturally channelled between these obstacles. Woods and villages may often be less than 1 km (0.62 miles) apart, so if you fire from the wood or village edge most of your targets are likely to be less than 500 m (1,640 ft) away.

Anti-tank weapons

In addition to your platoon medium anti-tank weapons — now the multi-purpose LAW 80, which has replaced both the 84-mm Carl Gustav recoilless gun and the disposable 66-mm M72 light anti-armour weapon in service — you may decide to take anti-tank mines, Claymore mines, phosphorus grenades, cratering charges or 'Molo-

49

MECHANISED COMBAT

tov' cocktails. In many armies the enemy formation you are likely to ambush will include motor rifle troops in APCs as well as tanks. In other theatres the column may include lorried infantry, towed artillery or logistic vehicles.

You will have a choice of mines. Current British mines are the Mk 7 anti-tank mine, the L3AI anti-tank mine (non-metallic), the L9AI bar mine and the off-route L14AI anti-tank mine.

Using anti-tank mines

The Mk 7 mine is a heavy metal mine containing 9 kg (19.8 lb) of explosive with an all-up weight of 14.5 kg (32 lb). You can't carry too many of these very far! However, it will cut the track of the heaviest-known tank and can be fitted with a booby trap. The mine is fired by the No. 5 Double Impulse Fuse, which will defeat tanks fitted with mine rollers; it will not operate under the weight of a man.

The Light L3AI non-metallic mine contains 6 kg (13 lb) of explosive and weighs 8 kg (17.5 lb). If you remove the metal detector ring the mine is not detectable by electronic detectors. The

Tactical use of the 66-mm LAW

The 66-mm Light Anti-Tank Weapon (LAW) is not capable of knocking out the latest generation of Soviet Main Battle Tanks or of dealing with a vehicle equipped with reactive armour. The British Army has replaced the LAW with LAW 80, a more accurate and more powerful weapon, but in many other forces you have to do the best you can with the American-made 66-mm.

It is perhaps most useful for 'bunker busting'. It has, in addition to a point contact fuse, a 'graze' fuse so that the round will detonate even without striking a target squarely. Even if you don't hit anything, the shock effect of the round bursting is considerable, and should give you the vital few seconds to get the enemy before he recovers enough to return fire.

A well-aimed shot from a 66-mm LAW can still achieve a 'mobility kill' against a tank by wrecking its tracks. LAW is also perfectly capable of taking out lightly-armoured vehicles like Soviet-built APCs.

TANK AMBUSHING

The British Army of the Rhine is deployed in an area dotted with large forestry blocks, towns and villages. You can force the enemy to deploy by engaging his armour in the open ground with long-range anti-tank systems sited in the woods and villages; in this close country you can do real damage with infantry anti-tank weapons, mines and demolition charges. Use obstacles and mines to channel the enemy into selecting lines of advance on which you have positioned your ambushes. Here you can hit the enemy hard on your own terms, withdraw, and hit him again. To do this successfully your tank ambush drills have to be slick.

Ambush positions
These should be manned by at least a section, and the 84 MAW and 66 LAW should be sited where they can fire into the side of the enemy tanks and be protected by Claymores and GPMG fire from infantry assault.

Protection
If you have time to dig in, do so, as this will protect you from return fire; and overhead cover is vital in woods if the enemy calls in artillery.

Siting
Choose an ambush position where the trees are close enough together to prevent movement by armour off the track you are ambushing.

Assess enemy strength
If their armour is supported by large quantities of infantry or if they get a chance to dismount in force from their BMPs, you could be in real trouble: set up the ambush so that the enemy can drive through. Here the advantages of command-detonated mines are obvious.

MECHANISED COMBAT

A sheep flees to cover as a 21-mm sub-calibre training round goes wide of the target on a firing range in Wales. Even against a static target, not firing back at 150 metres, it is not easy to achieve a hit with LAW.

The same target is hit with a live 66-mm round from a LAW: although its armour penetration is no longer sufficient to destroy the latest tanks, the warhead still has its uses.

The answer to a modern tank: volley-fired 66-mm LAW. If you each prepare two and fire a couple of synchronised volleys, the tank should be hit several times, hopefully forcing the crew to bail out.

Springing the ambush
To avoid early detection, create the road block as the ambush is sprung by destroying the lead vehicle by anti-tank fire or mines, or by using an explosive device placed in a culvert. You should destroy the rear vehicle at the same time.

Night action
Trip flares are very good if carefully sited; they can also be electrically detonated and used in clusters. Schermuly flares tend to give away the firer's position, and their use must be carefully co-ordinated with fire from the anti-armour weapons. Mortar and artillery illumination illuminate too great an area and can only be used in large-scale ambushes.

White phosphorus grenades
Excellent for creating a bit of mayhem. They also degrade night vision equipment and burn dismounting crews. This use of smoke is particularly useful, as motor rifle troops in BMPs are likely to be encountered.

Timed charges and Molotov cocktails
Prepare these well in advance.

Off-route mine
This is totally devastating against vehicles and effective against all WarPac main battle tanks. It can be initiated either by the target breaking a collapsing circuit or by means of a command wire.

Observation posts
These should be sited well forward to give advance warning of enemy approach, their strength and their direction of movement.

Road blocks
These can be created by felling suitable large trees or blowing craters. If they are not covered by fire, they should be mined and booby-trapped. Do not forget to mark them on the friendly side to prevent 'own goals'.

Mines
Decide where you are going to knock out the lead vehicle, and mine the area around it so that the following vehicles cannot drive round it out of the ambush. The same applies to the last vehicle: preferably you should choose a killing zone where there is no room to turn round.

MECHANISED COMBAT

Libyan BMPs lie abandoned in the Sahara desert after an ambush by troops loyal to President Habre of Chad. The area is being bombed by Libyan aircraft trying to destroy vehicles and equipment captured by Chadian forces. Moral: don't hang about at the site of an ambush.

WHITE LIGHT

The **Schermuly flare** (top left); the **pencil signal flare** (top right); and a **trip flare** (bottom). The Schermuly is a parachute flare that can give the enemy time to take cover as it flies up into the air. The trip flare provides instant white light at ground level, but must be carefully sited so that it illuminates the enemy, not the ambushers. Screens and mirrors can be used to increase its effect.

L3AI is only available in limited numbers and so is unlikely to be used for large defensive minefields, which makes it particularly suitable for laying by hand singly or in small numbers in a tank ambush. It will cut the track of the heaviest known tank.

The L9AI bar mine weighs 10.5 kg (23 lb) and contains 8.5 kg (18.7 lb) of explosive. It produces the same effect as a cylindrical mine, but with a lower density of mines (it is 1.2 m [4 ft] long and 0.1 m [0.33 ft] wide). It too will cut the track of the heaviest known tank.

Off-route mine

It is, though, the L14AI off-route anti-tank mine that is especially suitable for tank ambush; it can be set up quickly and covers a wide frontage. As its name implies, it is designed for use to one side of a track.

The off-route mine has a shaped-charge warhead, which is usually detonated when a tank runs over a triggering wire that has been stretched across the road. It will penetrate the side armour of any current tank, but if the copper slug penetrator hits the tank's road wheels rather than the side armour the tank may be immobilised – what is known as a mobility kill – instead of being completely destroyed.

It is usually set to fire horizontally at the passing target vehicle, and is set off when the vehicle ruptures the breakwire which is laid in its path. It is designed to penetrate up to 70 mm (2.75 in) of armour and has a maximum range of 80 m (262.5 ft), but it is at its most effective up to 40 m (131.2 ft). Its minimum range is 2 m (6.5 ft).

You can also site the mine to fire upwards, for example through a street manhole, to attack the belly of a tank; or downwards, for example from a window, to attack the top. The explosive is generally more effective against the top or belly of a tank than against its side, where road wheels and other appendages can reduce its effectiveness.

You will usually lay the mine so that it is set off by the target vehicle, but you can also set it up for command detonation, for instance by breaking the wire yourself at the required moment.

It will take you and only two other members of your patrol between five and 20 minutes, depending upon your skill and the nature of the site, to set up an off-route mine.

Value of grenades

You should also take phosphorus grenades and Claymore mines. The grenades will create confusion, degrade enemy night vision devices and cause burn injuries to dismounting crews. Site Claymore mines to kill dismounting troops and to protect the flanks of your ambush position against counter or surprise attack.

Having selected your mix of weapons, you now choose a suitable position for your ambush, from your personal knowledge of the battlefield, as a result of air reconnaissance, or just from having studied the ground on a map. Choose a defile or where the road passes between woods, large ditches, thick hedges, banks or buildings. If your ambush is slightly longer range and is crossing the open ground between two villages or woods, make sure that your wood frontage or village edge is defendable.

Springing the ambush

When the enemy tanks and APCs enter your ambush, attempt to disable the leading and rear vehicles. Site your MAWs and LAWs to engage targets from a flank. As your anti-tank projectiles slam into the sides of the enemy tanks, all hell will break loose. Some tanks will 'brew up', others will be immobilised, and some crews will attempt to leave their burning vehicles. Use your LSWs and Claymore mines to engage these crewmen and the supporting infantry who will debus in order to counter-attack your position.

Infantry threat

In close country it is not the tanks that will be the greatest threat to you, especially if you have prepared your ambush positions well by digging in: it is the enemy infantry who can get in amongst you who are the real threat. If you have chosen slightly rising ground and have the time to sow anti-personnel mines in front of your positions, you will be able to hold off a numerically larger enemy force for a surprisingly long time, long enough certainly to cause untold damage to his armoured vehicles. Relatively small numbers of Viet Cong ambushed US armoured columns in Vietnam with great success and then, at the right moment, vanished into the jungle.

Safe withdrawal

The timing of your withdrawal is likely to be crucial to the success of your operation. However many enemy tanks you destroy, your patrol will only be successful if you manage to extricate yourselves safely and with the minimum of casualties. You should therefore withdraw when the enemy is still surprised and shocked by your ambush, and before he starts to recover and reorganise.

The tank ambush is a classic infantry tactic. It relies upon stealth, cunning, surprise and shock action. It takes advantage of the main weaknesses of tanks. Above all, it provides the infantryman with the opportunity to meet strong enemy armoured columns on equal terms.

MECHANISED COMBAT

The BMD and the Blue Berets

Above: This BMD-2 command vehicle in Afghanistan has its antenna raised. Note the sixth road wheel and lack of turret. There is another version with AGS-17 30-mm grenade launcher.

Russian Airborne Forces (Vozdushov Desantniye Voyska, or VDV) have recently been transformed into powerful mechanised units, capable of seizing defended objectives and attacking well-armed forces deep in the enemy rear. Unlike British paratroopers, who drop with minimum arms and equipment and rely on speed and aggression to win the day, Russian airborne forces have a complete series of specialised vehicles that match all but the strongest allied forces in firepower and manoeuvrability.

Boyevaya Mashina Desantniye – the BMD

The BMD airborne amphibious infantry combat vehicle entered service in 1973 and at a stroke turned the VDV from light to mechanised infantry. Because of outward similarities, it is often mistaken for the BMP, but it has a totally new hull design and suspension, is far lighter, and is considerably more cramped inside.

The BMD can be carried in the hull of either the Il-76 'Candid' or An-12 'Cock' which, between them, account for the majority of Soviet military air transport. With its pneumatic suspension system folded up, the BMD can be dropped from the air.

The vehicle is pulled from the rear of the aircraft with the aid of a drogue chute. Then the main canopy deploys, and four probe poles unfold beneath the pallet. As soon as one of the poles makes contact with the ground, a retro-rocket system fires, considerably slowing the final stages of descent. The crew drop immediately after their vehicle and, during night drops, are guided to it by a radio 'bleeper'. They detach the pallet restraints and are operational in a matter of minutes.

The BMD's main armament is the same 73-mm low-pressure smoothbore gun as that mounted on the BMP. Although ineffective against the latest NATO tank armour, the high-explosive anti-tank (HEAT) round will penetrate the much thinner skins of Western infantry combat vehicles, but is only accurate to 800 m (2,620 ft). The BMD is itself very lightly protected (there is no more than 25 mm [1 in] of armour in the turret front and 15 mm [0.6 in] in the hull) – making it easy prey for the 25-mm (1-in) Chain Gun mounted on both the United States Bradley and the British Warrior.

Power and problems

A rear-mounted Type 5D-20 V-6 liquid-cooled diesel engine which develops 240 hp powers the BMD. It has a maximum road speed of 80 km/h (50 mph) and through water, aided by two water jets mounted in the rear, can reach 10 km/h (6.25 mph). The driver is seated centrally, just in front of the small one-man turret. Either the squad commander or gunner, seated to the left and right of the driver, can fire the section RPK machine-gun.

The main armament is unstabilised and, despite the help of a co-axial 7.62-mm PKT machine-gun, is very inaccurate when fired on the move. The automatic loader and 40 rounds of ready-to-use ammunition take up much of the remaining space inside, leaving room for only three passengers. Two AT-3 'Sagger' missiles are carried inside and can be fired from a rail above the gun barrel.

Despite its excellent reputation, the BMD does have a number of drawbacks. The 'Sagger' demands an unbroken line of sight between firer and target, and cannot be reloaded by the crew unless they break the vehicle's NBC seal.

The fuel tanks are poorly constructed and have a marked tendency to break away from their mountings, while the additional tanks in the rear are vulnerable to incendiary fire. Finally, the transmission is too fragile to withstand heavy drops – with the result that the shift lever can disengage at critical moments, leaving the vehicle helplessly stuck in gear.

Variations on the BMD

The most important, if least known, of the BMD family is the BMD-2 air assault transporter. First seen during the 1979 invasion of Afghanistan (and thus often called the BMD M1979 in the West), the BMD-2 is 60 cm (23.5 in) longer than the original, has an extra road wheel and return roller, and a built-up superstructure but no turret.

Two basic variants exist of which the BMD M1979/1 multi-purpose armoured transporter is the more common. Used for route control in larger drops and as a prime mover whenever support weapons such as the ZU-23 or Vasilek mortar are deployed, the transporter can also carry up to nine fully-equipped troops. Each soldier has a firing port, two capable of taking automatic weapons in the bow, two in the front hatches, two per side and one in the rear. The other variant, known as the BMD M1973/3 or BMD-2KSh, is a command vehicle equipped with a folding 'clothes line' antenna round the superstructure, a single commander's hatch and no firing ports.

MECHANISED COMBAT

The old ASU-57 self-propelled anti-tank gun on its pallet behind an Antonov An-12 'Cub' tactical transport. ASU-57s probably made their last combat appearance in the Ogaden desert during 1978 and are no longer in front-line service.

A BMD-1 in Afghanistan where Soviet airborne forces spearheaded the initial invasion in 1979 and later formed the cutting edge of the Soviet war effort against the Mujahideen. This BMD-1 appears to have lost its 'Sagger' launching rail.

There have been important changes in armament in a number of the newer BMDs. The 73-mm smooth-bore gun has been replaced by a far more accurate 30-mm autocannon (also mounted on the BMP-2), while the AT-4 'Spigot' has been mounted in place of the old AT-3 'Sagger'. 'Spigot' (known informally as 'Milanski', because of its marked similarity to MILAN) has a range of 2,000 m (6,560 ft) and can be taken from the vehicle and used on the ground. However, it can only be fired *from* the vehicle if a crewman opens one of the rear hatches and leans forwards. In this position, he is completely unprotected and would break the NBC seal if the NBC system were operating.

Fire support

Until 1985, Russian airborne troops were forced to rely on the D-30 122-mm towed field howitzer for artillery support, and on the ASU-85 for limited anti-tank protection. Whereas the D-30 was powerful enough to engage the majority of rear-echelon enemy artillery, it was too cumbersome to be towed by anything smaller than a BMD-2. The ASU-85, self-propelled, fast and manoeuvrable though it was, was far too small to engage modern MBTs.

The problem was partly resolved in 1985 by the introduction of the 2S9 assault artillery vehicle: a 120-mm breech-loading mortar, mounted in a large turret on a BMD-2 chassis. A small stub charge boosts the projectile out of the barrel, when a rocket motor cuts in to accelerate the round to cruise speed. Although not as accurate as the ground forces' 122-mm 2S1 howitzer, the 9-tonne (8.85-ton) 2S9 is light enough to be air-droppable. Its extremely high elevation lets it take on downhill targets (particularly important in Afghanistan), while in direct fire its HEAT round will penetrate all but the latest generation of tanks and will prove lethal when engaging APCs. Neither the range nor rate of fire are known, but experts believe that the mortar may well be fitted with a semi-automatic loading system, with burst-fire potential.

Air defence

A Russian airborne division relies on 48 SA-9 'Gaskin' surface-to-air missiles. Designed originally for low-level regimental defence, the SA-9 missile is derived from the SA-7 'Grail' but has a larger warhead, more powerful motor and improved controls. Carried in two double mounts on an adapted BRDM-2 vehicle, the missile is launched by its operator once he has acquired the target optically. Without up-to-date radar assistance 'Gaskin' can hope to have only very limited success against modern fast jets. But its usefulness against helicopters, probably the airborne soldier's greatest threat, is considerably greater.

Three SA-7 'Grail' hand-held launchers are issued to the air defence section of each airborne company. Perhaps the most famous system of its kind in the world, the basic missile consists of a tube with a dual-thrust solid motor steered by canard fins. Once the operator has located the target through his open sights, he simply takes the first pressure on the trigger,

Inside the BMD

The BMD APC is air-dropped as part of a Russian airborne operation and provides Russian paratroopers with increased mobility and firepower. It is seen here in action in Afghanistan.

15 mm hull armour
The thin frontal armour of the BMD is just enough to keep out small arms fire, but nothing larger.

2A28 73-mm smoothbore gun
Loaded automatically from a 40-round magazine, the 73-mm gun fires fin-stabilised, rocket-propelled HEAT and HE-FRAG rounds at up to eight rounds per minute. Its maximum range is 1,300 m (4,300 ft), but it is badly affected by wind.

7.62-mm PKT machine-gun

MECHANISED COMBAT

Guards of the Order of the Red Banner Airborne Division prepare pontoons for a river-crossing exercise. Russian airborne troops, like those of the Strategic Rocket Forces, all have pre-induction military training. No other branch of the Russian Armed Forces insists on this requirement, which is a measure of the importance attached to the seven paratroop divisions.

'Sagger' anti-tank missile
The early 1980s BMDs were usually seen with a 'Sagger' launcher above the 73-mm gun, as on the BMP. More recent pictures of BMDs show that many now carry a dismountable AT-4 'Spigot' missile.

Turret
The low turret is similar to that of the BMP. The armour is not more than 25-mm (1-in) thick over the frontal arc and two 'Sagger' missiles are stored within, plus 2,000 rounds for the co-axial 7.62-mm machine guns.

Driver
The driver sits centrally, observing through three periscopes positioned in front of his hatch. IR periscopes are available for night driving.

Gunner
The gunner relies on the familiar Russian stadiametric rangefinder, in which graticules coincide with different ranges assuming a target height of 2.7 m (9 ft; average NATO tank height). He keeps the controls for the 'Sagger' missile under his seat, pulling them out when the vehicle stops to fire a missile.

Commander
Sitting on the driver's left, the commander has access to the gyro-compass and radio.

Troop compartment
This is small, able to accommodate only three men in relative comfort, although more can be included. The only means of access is via a concertina-type hatch in the roof.

Periscope

Bow machine-gunner
Sitting behind and to the right of the driver, the bow machine-gunner operates the twin 7.62-mm PKT machine-guns.

Ground clearance
The independent suspension has a hydraulic mechanism for maintaining track tension and alternating ground clearance between 10 and 45 cm (3.9 and 17.7 in).

Idler

Road wheel

Track return roller

Drive sprocket

55

MECHANISED COMBAT

waits for the red light to turn green – indicating that the seeker has locked on – and applies full pressure to the trigger. The 2.5-kg (5.5-lb) warhead is lethal only against small targets, but can nevertheless force a large aircraft to abort its attack.

Infantry support

Among the latest, and finest, pieces of equipment to enter service with the VDV has been the AGS-17, or *Plamya*, which is now issued to the support section of each company. The AGS-17 is an automatic 30-mm grenade launcher; with maximum and effective ranges of 1,500 and 800 m (4,920 and 2,620 ft), respectively, and its staggering rate of fire of 65 missiles per minute, *Playma* would play a crucial role in suppressing enemy trench defences immediately before and during a final assault.

Each airborne soldier is armed with a 5.45-mm AKS-74 assault rifle capable of firing single rounds or automatic bursts and accurate out to 500 m (1,640 ft). It is obvious that Russian airborne units have a firepower second to none. Operating as totally self-contained entities, they can overcome enemy defences, defeat APC mounted and heliborne assaults, and defend themselves from armoured or airborne retaliation.

Above: First seen in public ten years ago, the SO-120 fire support vehicle carries an enlarged turret mounting a long-barrelled 120-mm mortar. This is replacing the ageing ASU-85 assault/anti-tank gun. The breech loading mortar is believed to carry a HEAT round for anti-tank action.

Below: Parading in Moscow, 7 November 1980, this BMD carries an AT-3 'Sagger' anti-tank guided missile on its launcher rail above the 73-mm smoothbore gun. The tiny size of the vehicle compared to the crew is readily apparent; the BMD is even more cramped than the BMP.

MECHANISED COMBAT

TANK HUNTING

Snarling, heavily-armoured and bristling with devastating armament, the Main Battle Tank can seem unstoppable to the infantryman — but there are occasions when determined and properly armed foot-soldiers can wreak havoc among even a sizeable detachment of enemy armour.

You'll usually carry out such operations at night, as part of a fighting patrol whose mission is to destroy enemy tanks at close range.

Known as tank hunting, this is a task for the ordinary infantryman, as distinct from the expert and highly-specialised techniques of anti-tank warfare. This section of Combat Skills details the weapons and techniques of tank hunting and tells you the vulnerable parts of an armoured vehicle that you should aim to hit.

Know the weak spots

The first thing you need to remember is that a tank is not by any means a flexible weapon. Closed down for combat, a tank crew has very limited vision. Close to, a tank is surrounded by blind spots.

TANK VULNERABILITY

1. Tanks have restricted vision when all their hatches are closed.
2. *Russian*-designed tanks in particular cannot depress their main armament very far, and are especially vulnerable when crossing a ridge.
3. The sides, rear and belly of a tank have much thinner armour than the front, which is usually proof against infantry anti-tank weapons.
4. Tanks are particularly vulnerable when re-fuelling and re-arming: they are stationary and bunched up.
5. Despite developments in thermal imagers, it is still difficult for a tank crew to spot infantry at night, especially if they use correct camouflage and concealment.
6. Tanks without infantry support are particularly vulnerable to infantry anti-tank weapons, especially in woods or built-up areas.

A Russian BMP Infantry Fighting Vehicle leads a mixed column of tanks and infantry. Successful tank hunting demands a thorough knowledge of the strengths and weaknesses of enemy armoured vehicles.

The tank's armament is virtually useless against a moving, close-range target, for the simple reason that it can't lower its gun sufficiently to engage a nearby target. As a result, a tank is at its most vulnerable when crossing a ridge — doubly so, in fact, because its tracks and lightly-armoured belly, rear and sides are also exposed then.

Tanks have to refuel and re-arm, usually at night. They'll do this either in a 'leaguer' (an administrative compound) or with a 'running replenishment' in the field. In either case enemy tanks will be bunched together, with little room to manoeuvre, and probably surrounded by cover that may screen them but also offers a perfect hiding-place for infantry. If you can find the enemy's tanks at such a time, they make excellent targets.

Imagine that you have been given the task of mounting a tank-hunting patrol. Your mission is to destroy as many tanks as possible as they undergo running replenishment in a

57

MECHANISED COMBAT

Taking aim with a Carl Gustav 84-mm recoilless rifle: this bulky but accurate weapon will penetrate the side and rear armour of modern tanks as long as they do not carry reactive armour panels.

village street just behind enemy lines. You know that the enemy has been using that location for two nights and that his tanks will be there again tonight. Recce patrols have already found a route for you between two forward enemy companies. You're to take a 12-man patrol.

As with any patrol, your job is to make sure you know as much as possible about the killing ground and your route to it. Use maps, aerial photographs and, if possible, survey the ground itself from an observation post.

Next, learn everything you can about the type of tank you are going to attack: where its most vulnerable points are, its hatches, radio antenna and sights. And make yourself familiar with enemy operating procedures.

Finally, rehearse your action on the objective and make sure that you and everyone going with you knows exactly what they have to do.

The weapons to take

Since your best chance of success lies in reaching your objective, making a quick attack and then withdrawing behind your own lines as fast and as unobtrusively as possible, you would ideally use timed charges. However, they are likely to be reserved for special forces.

A reasonable armoury of immediate-effect weapons for your 12-man patrol would include two 84-mm medium anti-tank weapons (MAWs), two 66-mm light anti-tank weapons

TANK HUNTING

Tank hunting is carried out by a fighting patrol which can vary in size from a section to several companies. The fighting patrol must have sufficient strength to carry out the mission and defend itself on route out and route back. The basic aim is to attack enemy tanks when they are harboured up, in a defensive position or in a tank leaguer.

Control, concealment and protection
As soon as you start firing your anti-tank weapons your position will be revealed to the enemy, so make sure you deploy these weapons at reasonable intervals, not all together.

Co-ordinate your fire
Make sure you can cover targets with both small-arms and anti-tank fire. A good co-ordinated fire plan will isolate individual tanks from their infantry support, restrict tank crews vision by forcing them to close down, and cut off the unit you are attacking from enemy reinforcements.

Withdrawal route
Make sure you have a secure withdrawal route, and leave a protection party back at the final RV. Everyone must know the signal for 'Break Contact' and what he must do. Withdrawal must be staggered, so that the enemy will not be keen to follow you up. Leapfrog back by half sections so that the enemy is always under fire while you withdraw.

Surprise
Your chances of success depend on catching the enemy unawares: your presence should be announced by a sudden hail of anti-tank and small arms fire. Your patrolling skills and fieldcraft will have to be excellent.

58

MECHANISED COMBAT

Information
Tank hunting relies on good intelligence information received in time to allow for careful planning and detailed preparation. Every eventuality must be covered.

Mines
Don't just use the 66-mm and 84-mm anti-tank weapons; plant anti-tank mines on likely enemy routes and approaches. Always mix some anti-personnel mines with the anti-tank mines to discourage the enemy engineers from digging them up.

Other weapons
Don't forget white phosphorus grenades: these are particularly effective against dismounted tank crew and add to the confusion. These grenades are also useful to cover your withdrawal, since they produce instant smoke and discourage the enemy from putting his head up.

ZSU 23-4
Remember, these self-propelled anti-aircraft guns can be used against ground targets with devastating effect. Get rid of them first! Even small-arms fire will wreck their optics and radar control system.

The 84-mm is a sizeable beast, but it needs to be in order to fire a large enough round to threaten a Main Battle Tank. In recent years tanks have received new types of armour designed to defeat infantry anti-tank weapons.

Timed charges and Molotov cocktails
Make sure these have been properly prepared before the patrol. Satchel charges should be double-fused to ensure that they detonate.

Know your enemy
You must be completely familiar with enemy AFVs and their tactics. You have to know what you are firing at.

(LAWs), both firing specialised high-explosive anti-tank (HEAT) missiles. The law is effective up to 200 m (656 ft), the MAW out to twice that distance. Both will penetrate the sides and rear of all but the latest Main Battle Tank. In addition, you'll be carrying phosphorus grenades, anti-tank mines and personal weapons.

You can attach image intensification (II) equipment or an individual weapon sight (IWS) to the MAW. This will give you an impressive picture of any nocturnal enemy activity at up to 150 m (492 ft) range even on the darkest night. Without an IWS, you will need white light – flares – to score a kill at anything but point-blank range.

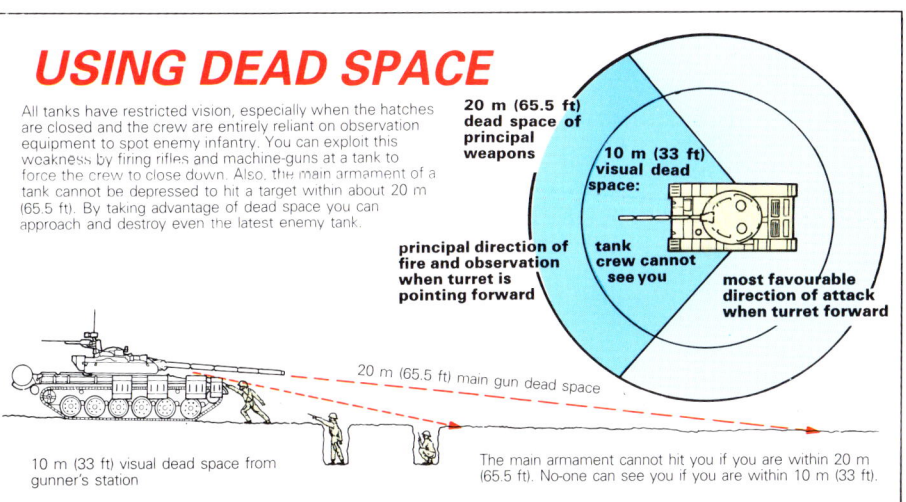

USING DEAD SPACE

All tanks have restricted vision, especially when the hatches are closed and the crew are entirely reliant on observation equipment to spot enemy infantry. You can exploit this weakness by firing rifles and machine-guns at a tank to force the crew to close down. Also, the main armament of a tank cannot be depressed to hit a target within about 20 m (65.5 ft). By taking advantage of dead space you can approach and destroy even the latest enemy tank.

20 m (65.5 ft) dead space of principal weapons

10 m (33 ft) visual dead space: tank crew cannot see you

principal direction of fire and observation when turret is pointing forward

most favourable direction of attack when turret forward

20 m (65.5 ft) main gun dead space

10 m (33 ft) visual dead space from gunner's station

The main armament cannot hit you if you are within 20 m (65.5 ft). No-one can see you if you are within 10 m (33 ft).

MECHANISED COMBAT

Into the attack

You should open fire with both MAWs and LAWs simultaneously. This creates the maximum immediate damage to the target and creates an invaluable psychological advantage.

Imagine the effect on the enemy: he thinks he is in a safe administrative area, behind his own lines, peaceably servicing his tanks. Then, several HEAT missiles slam into the vehicles.

Flames and panic break out everywhere. The tanks are waiting in column to be refuelled. Some reverse, others accelerate, some swerve to avoid other crippled and burning vehicles. At this point you should drop anti-tank mines at each end of the column, causing further damage and chaos. Meanwhile, rake any visible enemy personnel with automatic fire.

Above: When a tank comes over the crest of a hill it can expose its relatively thin belly armour to a well-sited infantry anti-tank weapon. Waiting behind the crest also shelters you from enemy fire until the last moment.

Above: This is the time to catch enemy armour – bunched up ready to refuel and rearm. A quick ambush with LAW 80 could inflict enormous damage.

About now some enemy crews will decide to abandon their ruined vehicles. At this stage, throw as many phosphorus grenades as you can into the midst of the tanks. This will disrupt any night vision equipment the enemy is trying to bring to bear on your position, and injure dismounting enemy troops. Having done your job, withdraw as discreetly as you can.

The ambush option

You can bring off an attack as successful as that only against an extremely inefficient enemy. But that is the ideal to aim for. In practice, most of your tank hunting will more nearly resemble an ambush.

If, for instance, you were withdrawing under the pressure of an enemy attack, one way of blunting his advance would be to set up tank ambushes and hit the enemy as he advances. You'll be able to mount these attacks most easily in close country or in built-up areas. If you can mount a series of ambushes, so much the better.

Choose your ambush site carefully. From it you should be able to get as close as 100 to 200 m (328 to 656 ft) to the enemy. Spring your ambush, withdraw along a pre-planned route, reorganise and repeat the process.

Any decently-trained tank crew will, if they can, avoid areas where they're vulnerable to ambush. And if they're forced to advance through close or built-up areas, they'll call on infantry to clear the woods or buildings first. However, in reality, tanks often outrun their infantry support – if they can get it at all. You will usually find rich pickings in a tank-hunting party working from a well-chosen ambush site.

Where to aim

The largest 'soft' area of a tank you're likely to see is its side. So wait until your target tank turns away from you before opening fire.

Tank tracks are particularly vulnerable. If you can blow off a track, the tank is as good as dead. It's immediately exposed as an easy kill for longer-range anti-tank missiles, and is costly for the enemy to recover.

The turret or the glacis plate at the front of the tank is where a tank's armour is hardest. Don't fire at the front of a tank, therefore, even if it's coming straight at you. Just take cover.

An infantryman hidden behind a building, an earth bank or in his trench is surprisingly safe from a tank. If one motors over your trench, just keep your head down. Then pop up behind it and send a LAW or MAW up its rear end.

Revenge is sweet

Tank hunting is a useful way for the infantryman to get back at the tank. It's not a practical way to destroy enemy armour in large numbers: this is a job for other tanks, long-range anti-tank missiles and anti-tank guided missiles (ATGW) on the open battlefield.

But tank hunting saps the enemy's morale and raises that of the foot-soldier no end – as well as reducing the threat to you from enemy armour. Tank hunting is the infantry's special contribution to the anti-tank battle.

When it's successful, tank hunting is truly an example of David slaying Goliath.

MECHANISED COMBAT

T-55 THE PEOPLE'S TANK

The T-54 series of Main Battle Tanks has seen more combat than any other type of post-war tank. They rumbled into Prague in 1968 to extinguish Czech hopes of freedom, and they crashed through the gates of the Presidential Palace in Saigon. They have fought in most African and Asian wars since the 1960s, and are still operated in large numbers amongst Russian-supplied forces worldwide.

Although the first T-54s were manufactured 40 years ago, the type is still operational all over the world, and various manufacturers are offering conversion deals to completely modernise this veteran tank.

The T-54 series includes a bewildering number of different models. The original production tank was soon altered, and many modifications introduced by subsequent versions were applied retrospectively, so it can be difficult to tell them apart.

Early model T-54s were very basic, World War II-style tanks, but had a very powerful armament for their time. The low turret presented an impressively small silhouette to the enemy, but only by cramming the commander, gunner and loader into

Below: A T-55 of the Russian Naval Infantry rumbles up a beach during exercises in the Baltic.

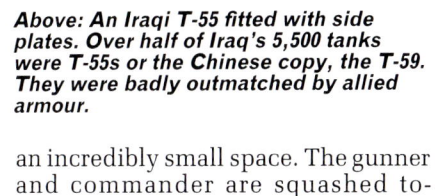

Above: An Iraqi T-55 fitted with side plates. Over half of Iraq's 5,500 tanks were T-55s or the Chinese copy, the T-59. They were badly outmatched by allied armour.

an incredibly small space. The gunner and commander are squashed together on the left of the gun while the loader sits to its right.

Three shells are positioned in a rack in the turret rear, and the loader has to twist round and load them into the breech with his left hand. They weigh 25 kg (55 lb) each, but his problems begin when these three have been fired because the T-54s and early T-55s have no turret floor. When the turret rotates, the gun breech moves round but you do not, unless you are sitting in your seat which hangs from the turret roof. If you are standing, you can be crushed by the breech.

The T-54/55's main armament is the D-10T 100-mm rifled gun, a mighty piece for its time but inadequate for the 1990s. The AP and HVAPDS-T ammunition for the D-10 lacks the penetration to destroy the latest generation of NATO armour except at very close range, and its HEAT round is equally ineffective. Against M48s, Leopard 1s and other lightly protected AFVs it will do the job, but the 15-20

MECHANISED COMBAT

Inside the T-55

This Syrian T-55 was knocked out on the Golan Heights during the 1973 Arab-Israeli war. Syrian numerical advantage was nullified by inflexible tactics, poor gunnery and lack of co-ordination.

Laser rangefinder
The weaknesses of the 100-mm gun were exacerbated on most versions of the T-54/55 a primitive fire control system. The Russians s no signs of re-arming th T-55s, but the addition c laser rangefinder greatly increases the effective range of the 100-mm g

Driver
The T-54/55 is a very uncomfortable vehicle to drive; double clutching to change gear and a dreadful suspension system make it an exhausting experience. The T-55s offered for sale by Israel have a much improved driver's station with steering wheel, new transmission and a decent suspension.

Hull front armour
This is 97 mm (3.8 in) thick, sloped at 58° on the top part, and 99 mm (3.9 in) thick on the lower part. The hull sides are protected by 79 mm (3 in) at 0° and the rear is just 46 mm thick.

Former East German T-54 training tanks in operation. Like many Warsaw Pact forces, the East German army continued to rely on the T-54/55 rather than convert entirely to the T-62. By 1989 the first-line units were equipped with T-72s, but T-54/55s still equipped second-line and training formations.

seconds needed to reload would cost it dearly in action against most NATO tanks with a competent crew.

Dangerous engine

The engines on Russian-built T-54/55s are often so badly made that they self-destruct: oil lines are easily blocked by a mass of loose metal filings and the engine overheats and catches fire. Made of magnesium alloy, it burns nicely and the whole tank can easily be destroyed. Some Warsaw Pact armies used to take the precaution of rebuilding the engines in T-54/55s they had bought, or, better still, manufacturing the engines themselves.

The T-54 was succeeded by the T-54A, B and C, each adding marginal improvements. The gun received a bore-evacuator, power elevation and both vertical and horizontal stabilisation: other changes included infra-red night vision equipment, automatic fire-extinguishers and an air filter.

T-55 introduction

The T-55 was first observed in the 1961 November Parade and introduced a more powerful engine, better transmission, a turret basket and nine more rounds of ammunition. Two years later, production switched to the T-55A, which featured an anti-radiation lining and an NBC system to protect the crew from nuclear fallout. This became the most numerically important model of the series, and many of its features were retrofitted to earlier tanks. The most significant improvement noted in recent years is a laser rangefinder, which began to appear on Warsaw Pact T-54/55s in the late 1970s.

Even with all these improvements, why is the T-54/55 still in such widespread service and important enough to warrant the expense of laser rangefinders? It remains a crude and simple combat vehicle harking back to World War II rather than looking forward to the 21st century.

Failure of T-62

One answer is that its replacement, the T-62, failed to live up to expectation: it is estimated that the T-62 cost about three times the price of a T-54/55 and it certainly did not provide triple value. Its armour protection is very similar, and its weight and battlefield mobility are also little different. The most significant change was the adoption of the 115-mm smoothbore gun which, although capable of destroying Western tanks at longer ranges, probably did not offer a wide enough margin to satisfy its critics. For these reasons the USSR continued to manufacture the tried and tested T-55 until 1981, a long time after it abandoned T-62 production.

China has copied the T-54 and produced one of the worst tanks inflicted on an army since 1945. Designated the T-59, the workmanship is appalling and it incorporates few of the improvements effected to the T-54/55s in Warsaw Pact service. The gun is not stabilised, so firing on the move is simply wasting ammunition; the choice of ammunition is more limited, and this is as badly manufactured as the rest of the tank.

The Pakistani army was supplied

MECHANISED COMBAT

Gunner
By modern standards, the armour-piercing shell from the T-54/55's 100-mm gun is severely lacking in penetrative power and its more lethal HEAT round is inaccurate. Worse, it takes 15-20 seconds to reload, since the gun must be fully elevated to give the loader room to extract the empty case and load a fresh round.

Commander
The commander has a cupola which can be traversed through 360° independently of the turret. On the Israeli-rebuilt T-55s the cupola is lowered flush with the turret.

12.7-mm anti-aircraft machine-gun
This appeared on later versions of the T-54 and is fitted to many of the T-54/55 variants.

This is the final production version of the T-55 with 'live' track, a laser rangefinder and radiation covers. Manufacture at the Omsk tank factory did not cease until 1981, and after 40 years in production about 50,000 are believed to have been manufactured in the USSR. When Chinese and Warsaw Pact production figures are added, the total number of T-54/55s built is near 100,000. Compare this with the production totals for its American contemporary, the M48 (11,703) or the British Centurion (4,423).

TPK-1 sight
This gives the commander reasonable visibility out to 400 m (1,310 ft) at night. The gunner's sights function out to about 800 m (2,625 ft).

Turret armour
The front of the turret is protected by 203 mm (8 in) of armour and the sides by 150 mm (5.9 in). This relatively thick armour combined with its low silhouette makes the T-54/55 well protected for a tank weighing only 36 tonnes (35 tons).

Loader
Only three rounds are stored in the turret, after which the loader must leave his seat to collect shells from the hull. T-54s and early T-55s had no turret floor, so if the turret rotated and the loader was standing on the hull floor he could be crushed by the breech.

Tracks
T-54/55s have Christie-type torsion bar suspension with five large road wheels. A useful recognition feature is the gap between the first and second wheels. The last production model shown here has 'live' track, but all previous versions had 'dead' (not under tension) track which hung loosely and sometimes led to the tank shedding its track.

with large numbers of T-59s and has been trying to rectify its shortcomings for many years. Ironically, their Indian opponents in the 1971 war used T-55s.

Russian T-55s which are still in service are being improved in several ways. This one carries a laser rangefinder fitted above the gun barrel.

Under new management

The Israeli army captured large numbers of T-54/55s in 1967 when the ill-starred Egyptian advance into the Sinai desert was decisively defeated. Many T-54/55s had broken down and were baking in the sun, waiting to be taken back for repair, when the Israelis overran them. Beggars can't be choosers, and the Israelis pressed them into service and used an entire brigade of them in the 1973 Yom Kippur war. That conflict netted another large haul of T-54/55s which were also modified for use against their former owners.

The Israelis extensively improved their T-54/55s by replacing the obsolete 100-mm gun with the L7 105-mm

MECHANISED COMBAT

weapon widely used in NATO, fitting a modern fire control system, new night vision aids and American machine-guns. From the point of view of the crew, the most popular improvement was the installation of air conditioning: a vital feature of a tank intended for desert warfare.

In the last few years Israel has offered modernised T-54/55s for export – tanks that bear little relation to the poorly made and badly maintained specimens originally acquired from the Arabs. Now powered by American General Motors engines with completely new transmission, they have a completely revised driver's station with a steering wheel instead of sticks and a good suspension system.

In the UK, Royal Ordnance has developed for the T-54/55 an update kit which centres around the replacement of the 100-mm gun with a British 105-mm weapon. This is specially aimed at the Egyptians and Pakistanis, who are keen to modernise their tank fleets. Egypt has also signed a deal with the American company, Teledyne, to bring their T-54/55s up to the Israeli standard.

Ageing vehicles

By the age of 40, most tanks are well into old age and are only fit for parades in African capitals. The T-54/55 was produced on a phenomenal scale; probably over 100,000 were manufactured, and the owners have a strong incentive to spin out the lives of these ageing vehicles. In several cases, notably Egypt and Libya, they form a major proportion of the army's tank strength, and modernisation is, of course, cheaper than buying new replacements.

The new generation of NATO armour – M1 Abrams, Leopard 2 and Challenger – are so cripplingly expensive that, outside the Alliance, only countries like Saudi Arabia can afford them; and the vehicles they replace, M60s, Leopard 1s and Chieftains, are unlikely to appear on the export market in major numbers. In the former Warsaw Pact, many armies never fully adopted the T-62 and units equipped with T-54/55s are moving directly to the T-72, their veteran tanks finally passing into reserve.

Israel captured many T-55s in 1967 and used them against their former owners in 1973. This Israeli T-55 carries a 105-mm rifled gun instead of the old Soviet 100-mm gun.

Above: Pakistani forces are mainly equipped with Chinese T-59s, which are cheap and cheerful copies of the Russian design. These are followed by a US-built M113 APC.

One of the more unusual T-55 variants offered by Western companies to Arab countries: a T-55 chassis fitted with Marksman twin 35-mm anti-aircraft guns.

An Iraqi T-55 booby-trapped by its crew before coalition forces overran the position. The 12.7-mm anti-aircraft gun has been removed from its mounting on the turret roof.

MECHANISED COMBAT

KILLING TANKS

5 POINTS FOR TANK-KILLING

1. Choose a firing position where you are firing into the side of enemy tanks but are protected and concealed from the direction of enemy approach.
2. Camouflage your position well, and use at least 45 cm of overhead protection over the weapon position.
3. Fire in support of the anti-tank weapons to your right and left; you should also be able to cover most, if not all, of the ground they cover.
4. Use mines and obstacles to channel the tanks into an area ideal for you to engage them in.
5. Choose your target carefully; concentrate on command tanks and SP AA guns, which provide air defence.

Enemy tanks, infantry fighting vehicles and other hostile armour represent probably the greatest threat you'll face when defending your position on the battlefield. This threat more than anything else will dictate how you set out your defences.

But preparing to defeat an armoured attack doesn't mean that you should slip into a defensive state of mind. Your tactics should be aggressive, imaginative and effective. This is when the enemy is at his most stretched and his most vulnerable – and you have a golden opportunity to inflict massive tank casualties on him.

Use natural obstacles to hinder and

A Russian-built T-62 emerges from a blazing forest, its stabilised 115-mm gun ready for the next target. As an infantryman the tank is still your most feared enemy, and unless you co-ordinate your anti-tank weapons it can smash your defences in minutes.

The Dragon system is currently in service with the US Army. The gunner just keeps the crosshairs of the sight on the target, and the tracker automatically guides the missile along the gunner's line of sight.

MECHANISED COMBAT

Above: The 84-mm Carl Gustav anti-tank weapon is recoilless, operated by two men and fires an 84-mm HEAT round. It has a considerable backblast signature and there is some doubt whether it can defeat modern Main Battle Tank frontal armour.

Right: The MILAN system replaces the ageing 120-mm Wombat; it will defeat all known threat armour out to a range of 1,950 m (6,398 ft). Again, it has a vicious backblast firing signature which affects survivability of the system.

Striker is a Spartan-armoured personnel carrier fitted with the Swingfire guided missile system. It will defeat any known armour combination from 150 to 4,000 m (492 to 13,123 ft) and is immune to electronic countermeasures.

impede the enemy, and to canalise his approach – that is, make him travel along the lines you want, to where you can ambush, harass or destroy him at will.

Remember that armoured vehicles are very limited by the ground that they can use. They need bridging or snorkelling equipment to cross anything but the smallest rivers or streams. Marshy or swampy ground is impassable to Main Battle Tanks, and close or wooded country, if not impassable, gives you an opportunity for tank ambush at close range.

Similarly, built-up areas delay and channel the movement of armoured vehicles and make them vulnerable to close-range infantry anti-tank weapons. You can of course thicken up all these natural anti-tank obstacles with minefields and, if you are defending a built-up area, with rubble, overturned buses and any other sort of artificial obstacle.

You can use 'dead' ground to conceal your defending, reserve and counter-attack forces. You can sight your anti-tank weapons in defilade positions (hidden from frontal observation) in order to provide enfilade fire (from a flank). You will then surprise the enemy from a flank and hit him where his armour is thinnest. The tank is also a bigger and easier target in enfilade.

You can also use reverse slopes. In other words, sight your anti-tank weapons several hundred yards back from the crest of a ridge or hill: your positions are then invisible to the enemy until he crosses the crest. You are safe from his long-range tank fire, but, as he shows his belly when he crosses the ridgeline, you can engage him with maximum effect. Clever use of ground is probably the most effective counter to the tank threat.

Exploit the conditions

You must exploit to the full any conditions that favour you. Despite the most modern night-vision and the most up-to-date thermal imaging equipment, tanks are more vulnerable in poor visibility. Finally, tanks do not like either close-country or built-up areas. Use these conditions when you can.

Well-planned and co-ordinated use of your anti-tank weapons will enable you to defeat enemy armour. In every battle group there is a combination of anti-armour systems available.

Infantrymen use hand-held missiles like the short-range, unguided LAW 80 and the wire-guided MILAN missile. Weapons mounted on infantry vehicles include medium-range missiles like MILAN and longer-range weapons like the American TOW. Modern IFVs and APCs are commonly armed with 25-mm or 30-mm cannon, that are able to penetrate light armoured opponents.

Heavy anti-armour systems make up the third category. These include Main Battle Tanks – the best killer of a tank is often another tank – and tank-destroyers armed with heavy missiles such as TOW, Swingfire or HOT,

BRITISH ARMY INFANTRY ANTI-TANK WEAPONS RANGES

This is the coverage of weapons you will use in an infantry battalion. The 66-mm LAW and 84-mm MAW are needed to cover the area between the firing line and Milan's minimum engagement.

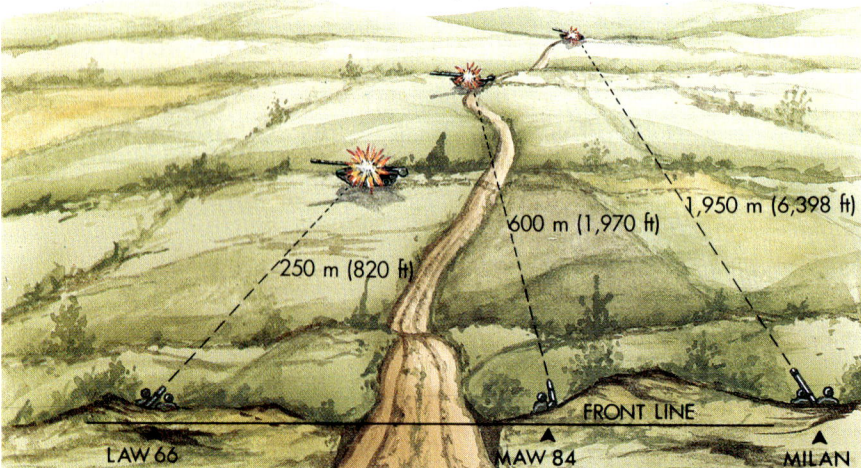

1. LAW can shoot out to 250 m (820 ft). 2. MAW sight range is 600 m (1,970 ft), but effective range is 500 m (1,640 ft) for static and 400 m (1,310 ft) for moving targets. 3. MILAN can shoot out to 1,950 m (6,398 ft). Missile flight time is 12.5 seconds; you must be able to track the target for the whole time.

MECHANISED COMBAT

which can take out enemy armour at ranges of 3 or 4 km (1.86 or 2.48 miles).

The helicopter has become a vital component in the armoured battle over the last few years. Machines like the US Army's AH-64 Apache are powerfully armed with up to 16 Hellfire missiles, and in recent conflicts the armed helicopter has shown itself to be amongst the most deadly of anti-tank weapons.

Minefields

Engineers are also important tank-killers, since they are responsible for anti-armour minefields. Even if they do not destroy or immobilise enemy tanks, mines cause delay at the very least. At best they will channel the enemy into a killing ground where other anti-tank weapons can be used.

Artillery bombardment

Artillery is also used in the anti-armour battle. A direct hit from a gun of 155-mm or above is enough to shatter the most heavily armoured of tanks. But the guns don't have to hit to be effective against a massed tank attack. A concentrated artillery bombardment can ruin optics, destroy radio antennas, dislodge and set fire to external fuel tanks and disorient and disconcert tank crews. Multi-barrelled rocket systems such as MLRS can fire rockets that scatter bomblets designed to penetrate the weaker top armour of tanks. Ground attack aircraft such as the Harrier and A-10 are most effective tank destroyers: they are capable of either rocket or bomb attack against tank targets.

You will see from this brief gallop through the systems available to you, or in direct support of your battalion, that there is a vast array of weapon systems capable of defeating a tank attack. It is precisely because there are so many systems that they must be carefully co-ordinated in order to avoid duplication and waste.

Hand-held missiles

Closest to you will be the hand-held weapons. Guided missiles are designed for use at ranges up to 2,000 m (6,562 ft). Light anti-armour weapons are effective at 500 m (1,640 ft) or less. Missiles like MILAN can be fitted with a thermal imaging system, so that you can use them 24 hours a day and in bad weather. MILAN and the short-range systems are complementary. You can use them to fill in gaps in the

FIGHTING THE ANTI-ARMOUR BATTLE

Tanks combine firepower, mobility and armoured protection to produce what is known as 'shock action'. In the past, the quality and quantity of Warsaw Pact armour, combined with a massive indirect firepower capability, formed a serious threat, and so all defence on the NATO central front was designed around the anti-armour plan.

Chieftain
In positional defence, it is normal to fight in mixed teams of tanks and infantry with dedicated artillery support, as well as some signals and engineers. If you have tanks with you, then make sure the anti-tank plan is co-ordinated to include them. Each tank will have a number of fire positions pre-prepared, the idea being to fire two or three shots and move; this will give the tank a better chance of survival.

Small-arms fire
7.62-mm rifle and GPMG fire will force tanks to close down, making target acquisition more difficult. 0.50 calibre rounds will damage the BTR-60 and similar vehicles.

Milan
Quick to deploy and devastating in positional defence, Milan would normally be fully dug in. The long flight time of the missile means you must be able to see the target for a full 12.5 seconds at maximum range, and the missile can be decoyed by other infra-red sources on the battlefield, like burning tank hulks. Milan must be deployed with short-range anti-tank protection (84-mm) and with infantry in a position to defend it.

Weapons siting
In anti-tank warfare your siting of weapons is all-important to minimise your vulnerability to the enemy's direct and indirect fire.

MECHANISED COMBAT

Above: The 66-mm LAW (Light Anti-tank Weapon) is a one-shot, recoilless, throwaway weapon, not particularly accurate and not capable of defeating modern MBT armour, though it has its uses for APCs and bunkers.

Below: Indirect artillery fire will not generally knock out Main Battle Tanks; however, it will cause them to 'close down', restricting their field of view, and fragments will smash optics and radio aerials and damage tracks and running gear.

Above: A light tank like the Scimitar is not designed to fight but to recce; the 30-mm RARDEN cannon could damage a T-72 but certainly not knock it out. However, it would be very useful on BMPs and other APCs and thin-skinned vehicles.

MILAN defence, or you can use them to provide close anti-tank protection for isolated MILAN crews or at distances below MILAN's minimum range.

In a mechanised battalion you will have your vehicles near you in your defensive position. Site them so that you can use their weapon systems to best advantage. Use the cannon mounted on APCs and reconnaissance vehicles to engage enemy APCs, and other lightly armoured vehicles.

When you operate in a mechanised battle group you will be supported by tanks. The tank is the most effective tank-killer of all. It can fire its armour-piercing discarding sabot (APDS) rounds out to 2,000 m (6,562 ft) with great accuracy, and at a rate of up to eight rounds a minute.

Use of tanks

However, tanks are best used to achieve surprise. Modern armies rarely use tanks as static gun platforms. That would be a waste of their mobility. They are kept in reserve, ready to cut off and destroy any enemy tank penetration.

The next component of your anti-tank plan is the minefield. This is a subject in itself; at this stage all you need to know is that the anti-tank mine plays an important part in the overall plan to defeat an enemy armoured attack.

There are several categories of anti-tank mine: the most common are the conventional cylindrical pressure mine (such as the British Mk 7), the bar mine, the off-route mine (designed to attack the side of a tank) and the scatterable mine, which can be fired from a gun or launched from a system mounted on an APC. Well-planned minefields covered by fire from your defensive positions can cause havoc among an enemy armoured formation.

Anti-tank helicopters

Anti-tank helicopters are also a subject in themselves. TOW missiles fired from Lynx have a range out to 3,750 m (12,300 ft). They are likely to engage massed enemy tank attacks of over 60 armoured vehicles well out to the front of you. Your role will be to mop up what is left.

Add to this array of weapon systems the anti-tank capabilities of both artillery and offensive air support, and you will see that you stand a very good chance of blunting, stopping and destroying even the most concerted armoured threat. The tank is still a potent weapon system but it is no longer queen of the battlefield: the armoured helicopter is emerging as a contender for that title.

US ARMY INFANTRY ANTI-TANK WEAPONS RANGES

This shows the corresponding coverage for the US Army. TOW gives one extra km (0.6 mile) over MILAN, but there are not that many positions that will allow a clear shoot-out to that range. Dragon again has better maximum range that MAW, but cannot be fired as rapidly.

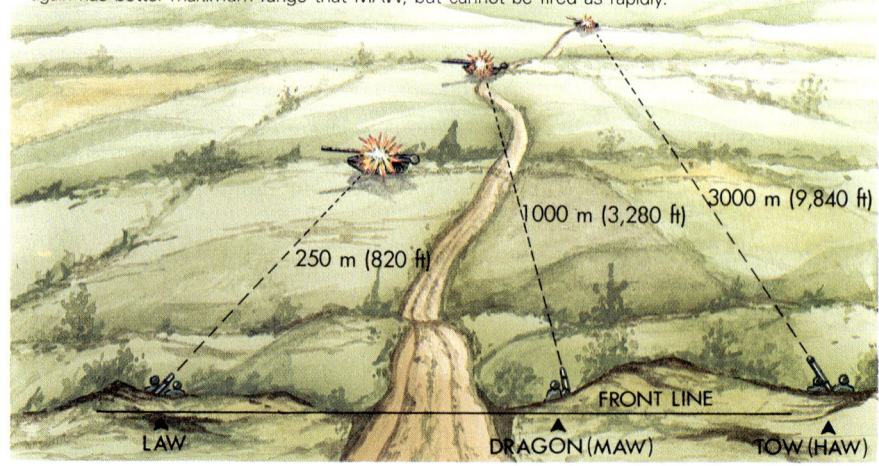

1. LAW: a more realistic range for engaging a tank would be 150 m (492 ft) or less, and volley fire is recommended.
2. Dragon can shoot out to 1,000 m (3,280 ft), and has a minimum range of 65 m (213 ft). The gap is covered by LAW.
3. TOW can shoot out to 3,000 m (9,840 ft) and must be carefully sighted to take advantage of this.

MECHANISED COMBAT

ON GUARD WITH RAPIER

Rapier streaks away at Mach 2 in pursuit of an enemy aircraft. When it hits, the direct-action fuse explodes the warhead inside the target, with catastrophic results.

A combination of radar, television, micro-processors and two highly-trained soldiers bring the awesome killing power of the Rapier surface-to-air missile system to bear on a hapless enemy aircraft. Designed as a fast, mobile defence against low-flying hostiles, the deadly accurate British Aerospace Rapier is now in service with the armed forces of 14 countries.

The most basic form of Rapier consists of four missiles mounted on a launching platform that can turn through 360 degrees, a surveillance radar system and command transmit-

You can re-load Rapier missiles quickly and easily without cranes or special kit. Once they are loaded the system can be made ready for firing in under 15 seconds.

MECHANISED COMBAT

Inside Rapier

Command aerial — This provides the link between the missile in flight and the computer in the base unit which controls its flight.

Missile — Rapier is designed to destroy low-flying aircraft or helicopters, and the fast reaction time of the system enables it to engage targets which suddenly emerge from behind cover.

RAPIER MISSILE IN DETAIL

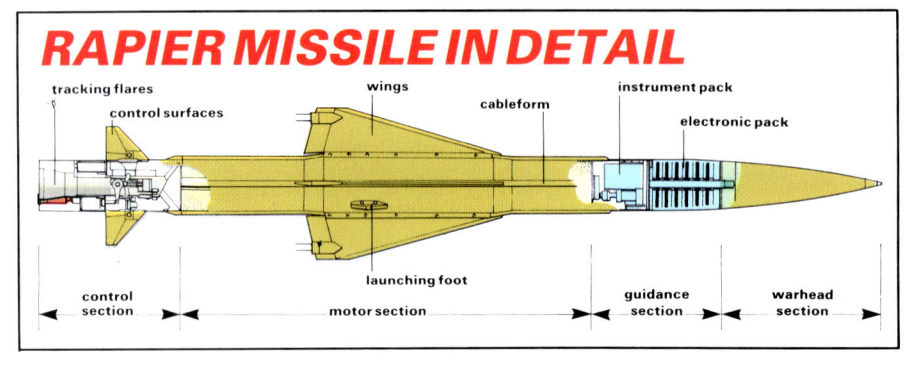

The Rapier missile's streamlined body is divided into four sections: the warhead, the guidance system, the propulsion unit and the control section. The nose is made of collapsible plastic moulded to the optimum aerodynamic shape. Rapier requires no maintenance or servicing and has a shelf life of over 10 years.

ter, an optical tracker, and a power generator.

The skeleton crew consists of two men: a radar operator and a tracker operator who keeps the target in view through an optical binocular sight. A mini-computer housed in the base of the launcher co-ordinates the weapon and the tracker operator.

Computerised killer

When the radar picks up a potential target it electronically 'interrogates' the aircraft for its Identification, Friend or Foe (IFF) code. If the aircraft is not emitting the appropriate signals declaring it to be friendly, the radar swings the launcher to the correct firing angle, tracks the target until it is 'in cover' – that is, within range – and gives the signal to fire.

The missile is automatically aligned to fly into your line of sight when you launch it. You continue to track the target through the optical sight, while a TV camera on the launcher monitors the path of the missile by tracking the flares in its tail.

The camera is linked to the fire unit computer, which compares the course of the missile with the course of the target. The computer is connected to a transmitter which sends signals to the missile whenever it needs to change course.

Blindfire

Obviously the system will only work if you know your job. It assumes that your optical tracking sight is pointed at the target and will send the missile wherever you are pointing it. At night or in bad weather you have little chance of visually tracking an enemy jet, so British Aerospace and Marconi Space and Defence Systems

The tactical control unit inside Tracked Rapier provides tactical control: its panel is divided into 32 sectors each of 11¼°. You can give one sector priority or exclude it from operation if necessary.

The tracker operator's station inside Tracked Rapier allows him to swivel through 90° to face either the tracker controls or the front of the vehicle where he can operate the radios.

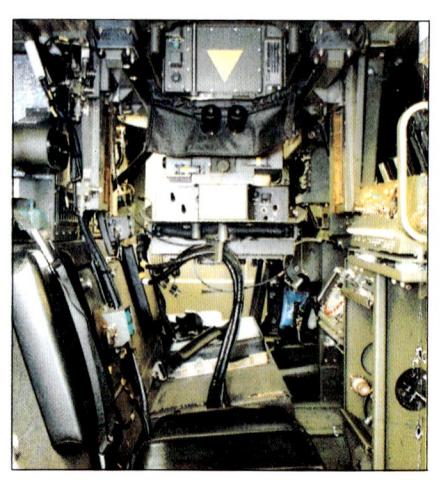

MECHANISED COMBAT

Surveillance radar and IFF
Rotating once every second, this alerts the crew to the presence of enemy aircraft. Its relatively low silhouette increases Tracked Rapier's battlefield survivability.

Armoured missile bin
Each contains four Rapier missiles. All eight missiles can be re-loaded by two men in just under five minutes.

Thermally Enhanced Optical Tracker (TOTE)
To provide day and night capability, the operator is provided with an optical tracking system and a thermal imaging tracker. All he has to do is keep the target within the crosshairs of his binocular sight and the Rapier system will send a 42-kg (92.5 lb) missile directly into the target aircraft.

have designed Blindfire. This uses radar to track the target and guide the missile.

On its own

Blindfire is mounted on the same chassis as the towed Rapier system and has its own generator. When you are using Blindfire the system practically works on its own: as soon as the surveillance radar detects enemy aircraft and sounds the alarm you select 'Radar'. Blindfire's tracking radar immediately swings round to the correct angle and elevation. As soon as it has 'locked on' to the target and you receive the 'In Cover' signal, you hit the fire button and launch the missile.

Everything is now automatic; the radar tracks both the target and the missile, and the fire unit computer modifies the missile's course as the target manoeuvres to escape destruction.

Unless you launched the missile at near maximum range the target will have a tough time outrunning Rapier, which maintains a speed of Mach 2 out to 7,000 m (22,965 ft), and no aircraft can out-turn it. Unless the enemy aircraft can escape behind cover, such as a large hill, it is unlikely to survive the experience.

In the hands of a skilled operator, Rapier missiles rarely miss – which earned the system the nickname 'hittile'. The average kill rate is over 70 per cent. And when the missile does make contact, the effect is devastating.

The aerodynamically-shaped plastic nose cone breaks on hitting the target, so that the hardened head smashes through the structure before setting off the aft-mounted crush fuse. The result is that the HE warhead detonates *inside* the aircraft, with predictably shattering consequences.

In the field

In British service Rapier travels in two guises. One is known as Towed Rapier, operating in a mobile optical fire unit (MOFU) that is based on two Land Rovers. One vehicle carries two men, four missiles and the optical tracker, and tows the launcher unit. The other vehicle carries three men, stores and supplies, and tows an additional nine missiles.

A MOFU can operate indepen-

MECHANISED COMBAT

Towed Rapier was deployed in the Falklands as part of the islands' modern defences. A large number of Rapiers were fired during the Falklands war, forcing Argentine aircraft to fly very low and fast. At least one Dagger jet was destroyed by a direct hit.

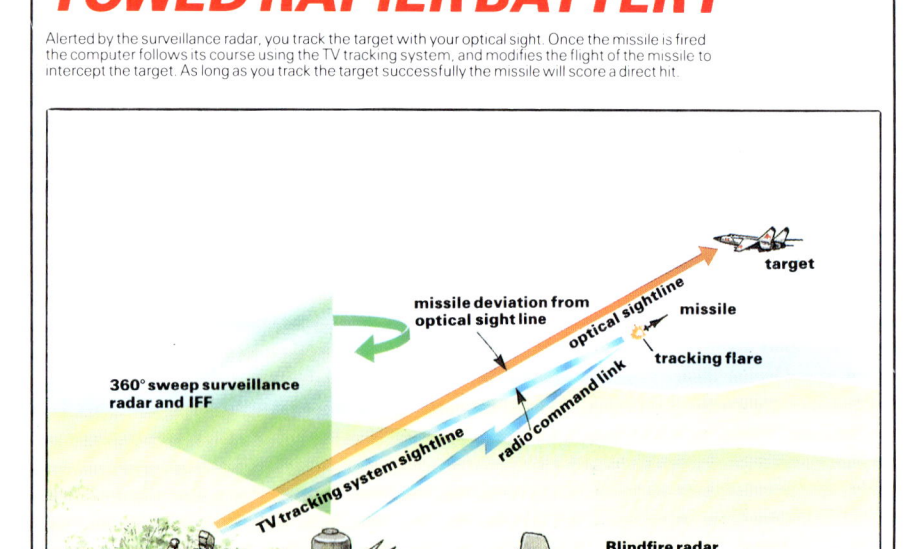

*Tracked Rapier is highly mobile and can accompany tanks over the roughest ground. It is amphibious, air transportable and can be operated by crew members wearing **NBC** protective kit.*

TOWED RAPIER BATTERY

Alerted by the surveillance radar, you track the target with your optical sight. Once the missile is fired the computer follows its course using the TV tracking system, and modifies the flight of the missile to intercept the target. As long as you track the target successfully the missile will score a direct hit.

- 360° sweep surveillance radar and IFF
- optical tracker
- power supply units
- launcher
- **Blindfire radar** — This tracks the target and guides the missile, allowing Rapier to be used at night or in fog.
- missile deviation from optical sight line
- optical sightline
- missile
- tracking flare
- radio command link
- TV tracking system sightline
- target

dently or as part of a larger, co-ordinated air defence deployment. It can be shifted very quickly into action as the whole unit is light enough to be moved by either fixed-wing aircraft or helicopter.

Tracked Rapier is the system's second form, developed to give great mobility in battlefield conditions and extra protection to the crews. This is a combination of the Rapier launcher, missiles and a specially-adapted RCM 748 variant of a US M548 tracked cargo carrier.

The British Army's Tracked Rapier launchers have been improved by the addition of TOTE (Tracked Optically Thermally Enhanced) equipment: an infra-red telescope scanner and an electronics system linking it to the existing optical tracker. This tracks the heat emissions from enemy aircraft, allowing you to operate at night or in poor visibility.

The Rapier system has been continually updated, although the missile itself has remained largely unaltered as its aerodynamic performance is excellent.

More work by British Aerospace has further reduced the amount of work demanded of the operator: 'Laserfire' is a new version of Rapier using an automatic laser tracker and a new surveillance radar. This has high immunity from enemy electronic counter measures and provides a very cost-effective night and day anti-aircraft system.

The British Army has adopted another development, the Rapier 2000, during the 1990s. This combines tracking and launch radars with optical and infra-red tracking, while the missile has a proximity fuse enabling it to destroy small targets like cruise missiles.

MECHANISED COMBAT

COUNTER-PENETRATION: PLUGGING THE GAP

Counter-penetration: blocking an enemy breakthrough (or 'penetration') by deploying troops on the ground to stop the enemy and set him up for a counter-attack. There is always the danger in battle that the enemy will concentrate overwhelming force at one point on his front, and so achieve a breakthrough. The defender, however, has to *spread* his forces, so that he is ready to meet the enemy wherever he begins his main thrust. It is, therefore, almost impossible to prevent a breakthrough at some point.

The secret of winning the battle in the end is to have sufficient mobile reserves to let you react quickly to a breakthrough and plug the gap. The force that plugs the gap (or counter-penetrates) may not be able to destroy the enemy force that has broken through. If it can, so much the better. But if it can't, it does not necessarily matter. Its main purpose is simply to stop the enemy, make him deploy and force him to mount an attack. This will buy time for your armoured reserves to redeploy and mount a counter-stroke.

One way to deal with this is to mount counterstrokes into the flanks of these enemy wedges. But this may not be enough. The enemy will defend their flanks against counter-attack. Armies need forces that are capable of blunting and stopping the head of a breakthrough.

These forces do exist. They are either airmobile (moved into position by helicopter) or equipped with wheeled vehicles so that they can move rapidly to the area where they're needed. The Americans have the most powerful of all these forces. It is called 6CBAC (Combat Brigade Air Cavalry), and is equipped with a very high proportion of attack helicopters with a lethal anti-tank capability. In the right place, at the right time, it can cut a tank division to pieces in one devastating attack.

The Germans have a similar airmobile capability, while the Belgians and Dutch have wheeled battalions that are trained in counter-penetration. The British have a highly-trained and extremely effective airmobile brigade, stationed in England, which was

A US Army TOW missile team in position overlooks a wide valley with 'killing zone' written all over it. NATO plans for defending Europe rely heavily on mobile units of anti-tank guided missiles to whittle down enemy tank strength. The British Army now has an airmobile brigade equipped with MILAN, which would be flown to the crisis point by helicopter.

BUYING TIME

For 40 years NATO's nightmare was a Soviet attack on Western Europe, spearheaded by powerful armoured forces driving deep into Germany. NATO's defensive plan was to:

1. **Attack the flanks of the Soviet columns.**
2. **Deploy fast reaction units in the path of the Soviet forces to attack their tank strength.**
3. **Counter-attack the weakened Soviet spearheads and defeat them with NATO armour.**

73

MECHANISED COMBAT

Above: A Chieftain tank and two FV 432 APCs advance across open country covered by another tank waiting in the tree line.

A Chieftain ARV heads down a forest track on a pre-planned withdrawal route. Following enemy forces will run into hastily-laid bar mines and a series of ambushes.

earmarked as a reinforcement brigade for BAOR. Its main role was counter-penetration. It is also equipped with wheeled vehicles so that, if weather prevents flying, it can still deploy quickly by road. But its main means of moving soldiers quickly over long distances are support helicopters, Pumas and Chinooks crewed by the RAF. Large numbers of these are deployed in Germany, and they can deploy the brigade, its weapons, its ammunition and some of its vehicles, in a matter of hours – the exact time depending on the distance to be flown and the number of helicopters available.

The tactics you'll use

From a study of the ground and the routes available to the enemy it will be obvious which route he will have to take. For example, the terrain in Ger-

MECHANISED COMBAT

many is full of choke points through which the enemy will inevitably have to pass – defiles, gaps, valleys, bridges and so on – and it is on this kind of position that you will counter-penetrate. Your commander will decide exactly where you should stop him, and aim to put you in position some hours before the enemy is due to arrive. A reconnaissance party will have flown in ahead of you and planned the defence.

You will have all the advantages that the ground can offer, and you'll have the weapons to do the job. First, you must dig in and prepare your defences and fields of fire as best you can in the time available. The battalions in your brigade are equipped with all the normal infantry weapons, but what makes them different is that they have 42 MILAN firing posts.

This gives you a stupendous tank-busting and tank-stopping capability. It means that you, in an infantry battalion, really *can* stop an enemy tank regiment.

The only thing that you will lack, dug in, is mobility. Once your support helicopters have dropped you, you're committed to fighting in your prepared position. But mobile firepower is on hand, from Lynx anti-tank helicopters, with eight TOW missiles (and eight more in reserve), a range of 3,000 m (9,840 ft), and even greater lethality that MILAN.

When you 'fix' the enemy by stopping him in front of your position, the Lynx attack helicopters will come sweeping in from a flank to deliver a devastating blow against the enemy's armour – which you will have bunched up in front of you, making an ideal target. With luck, you'll also have support from Harriers and A-10s.

Above: When the path of the enemy thrust has been identified and units assigned to a counter-penetration operation, the flags on the staff maps will all be in position.

TOW-armed Lynxes provide massive anti-tank firepower, and the combination of American A-10s and AH-64s could very well be decisive. Woods, built-up areas and other obstacles will also channel the movement of enemy armoured forces, where infantry armed with MILAN can inflict serious losses.

Below: A jeep-mounted TOW missile blasts off down range. These unprotected missile mounts are good for the single shot scenario, but cannot survive the hail of suppressive fire a missile launcher must expect in a protracted action.

75

MECHANISED COMBAT

You won't win a war by counter-penetrating. It's a necessary stop-gap against an enemy who has the initiative. Counter-penetration is like blocking your opponent's punch with your left hand, to set him up for a swinging blow with your right hand. Although your own anti-tank helicopters may be able to provide that knockout blow, it's more likely that you'll need a heavier punch in the form of armoured reserves to defeat an enemy breakthrough finally. But it takes time for an armoured brigade or division to get to the scene of the breakthrough. An airmobile counter-penetration force gains that time by staying in reserve well behind the front line, and then flying in ahead of the enemy to stop him in his tracks.

Use your ground

A counter-penetration force does have limitations. It's no good putting a brigade in front of a massive tank advance in the middle of a flat open plain. You will dig in, the enemy will run up against you, discover that you have nothing in either flank, and simply bypass you.

A counter-penetration force is effective only if it is blocking a gap through which the enemy must travel with armoured forces. In northern Germany, the terrain is interlaced with gaps, defiles and valleys that provide 'shoulders' for your positions. Of course, the enemy can try to use dismounted infantry to get at you on these 'shoulders', but it will be difficult, inconvenient and time-consuming. Meanwhile, his tanks will be bottled up in a choke point and extremely vulnerable to attack. Most important of all, he will have lost momentum – and maintaining momentum is all-important for a successful breakthrough. Your main role, remember, will be to gain time.

Getting out

Assuming you're successful, and that your brigade defeats a penetration, you can be quickly extricated by helicopter and redeployed to meet the next threat. The Air Mobile Brigade is the most flexible instrument of firepower in the British Army. Although lightly armed compared with an armoured brigade, its extra complement of MILAN, its integral 81-mm mortar platoons and the supporting Lynx TOW helicopters make it strong enough to do the job.

Its unique characteristic is that, within a few hours of a threat being identified, the Chinook and Puma helicopters dedicated to the brigade will have lifted the entire force into position.

A normal formation does not have this ability. It takes days rather than hours to move an armoured brigade any great distance. While on the move it is vulnerable to air attack, which invariably means it must move by night. Of course, any kind of formation can occupy a counter-penetration position. It doesn't *have* to be airmobile, but the organisation and capability of an airmobile brigade makes it particularly suitable for this task.

When your helicopter drops you at your counter-penetration position, get out your spade and get digging! Before you know it, you may have a tank division staring you in the face. You've got to be able to take him on, stop him and set him up for the kill.

Above: Former Chieftains of the Royal Scots Dragoon Guards on manoeuvres. Only a full-scale counter-attack by NATO armour would have been enough to smash a Soviet Operational Manoeuvre Group on its way to the Rhine. All Chieftains have now been replaced by Challengers.

Light and low but heavily armed, former Soviet tanks were ideal for combat on open plains, but if they had invaded Germany they would have been on infantry territory, with few stretches of land devoid of forest or built-up areas.

Right: Identifying the direction of the main enemy thrust will be a difficult job. CVR(T)s like this Scimitar will play a vital role, fighting for reconnaissance information against the opposing armoured screen.

The anti-tank version of the Spartan APC series is armed with MILAN – two missiles ready to fire and another eight inside. It will have to move rapidly to new firing positions to avoid being knocked out by enemy tank guns.

MECHANISED COMBAT

RATEL
Bush Fighter

A column of Ratels moves through the gathering dusk as the South African Defence Force goes onto the offensive against SWAPO guerrillas in Angola. At this time, in the early 1980s, the SADF was the most experienced bush-fighting force in the world.

The Ratel is tough — it has to be. The terrain it operates in is some of the most hostile in the world, which alone inflicts harsh punishment, to say nothing of the guerrillas against whom it is deployed. The Ratel infantry combat vehicle is named after a species of South African honey badger. Despite its small size, the original Ratel is a fierce creature that is not only capable of absorbing a great deal of physical damage but is an implacable and aggressive fighter. The Ratel vehicle is thus well named, for it too is a formidable opponent.

Tough conditions

The first Ratel was built in 1974 and was designed from the start to be used under South African conditions. This means long missions over rugged and variable terrain where it can expect little maintenance. It also has to be well armed and able to carry all its own supplies, spares and a good-sized complement of men.

Over the years the Ratel has met all these demands and there are now several versions of the basic vehicle, all of them using the same basic 6×6 drive and armoured hull. The powerpack is a six-cylinder diesel engine located at the left rear. It provides the power needed to drive the Ratel over all manner of harsh terrain and the punch to push the vehicle through dense bush, including young trees.

All-terrain vehicle

The Ratel is a true all-terrain vehicle. It rarely has to cross water obstacles under South African conditions, but if

The Ratel was designed for speed and long-range operations over bush terrain. It proved more than capable of dealing with such rough handling.

it does, it can ford up to 1.2 m (4 ft). From his central position which is well forward in the hull the driver has excellent vision through three large bullet-proof windscreens. These are covered by armoured shields if there is any serious shooting in progress, and he then uses periscopes. The gear-

MECHANISED COMBAT

box is automatic and the vehicle is very easy to drive.

Blast protection

The driver has his own roof hatch but he can enter and leave his position from within the vehicle, so if he is wounded he can be replaced without anyone having to leave the shelter of the armoured hull. The hull is designed to be proof against the effects of land mines: any that go off under the hull will just blow off wheels and will only rarely penetrate the shaped armour, which has been specially designed to withstand blasts from most anti-tank mines.

If a Ratel is damaged on operations it is never left behind for an enemy to loot. Every Ratel can be fitted with a small crane jib or tow bar at the rear, and these are used to tow damaged vehicles out of action.

Hydraulic doors

The hull has entry doors at the left and rear, operated by hydraulics to ensure they open and close at all vehicle angles. There are also many roof hatches, often left open most of the time until the shooting starts. A Ratel can carry up to 11 fully-armed and equipped troops, including the driver.

For a vehicle of its type the Ratel is very well armed. The basic vehicle, the Ratel 20, has a two-man turret mounting a 20-mm cannon and a 7.62-mm MG4 co-axial machine-gun, the South African version of the old 0.30-in Browning M1919. Over a hatch at the rear there is another MG4, used for local and air defence, while some vehicles also carry an extra MG4 over the turret. The occupants can fire their R4 rifles through the four firing and vision ports provided along each side. South African soldiers frequently supplement their armament with captured RPG-7 rocket launchers, and the Ratel is roomy enough to accommodate them.

Inside the Ratel

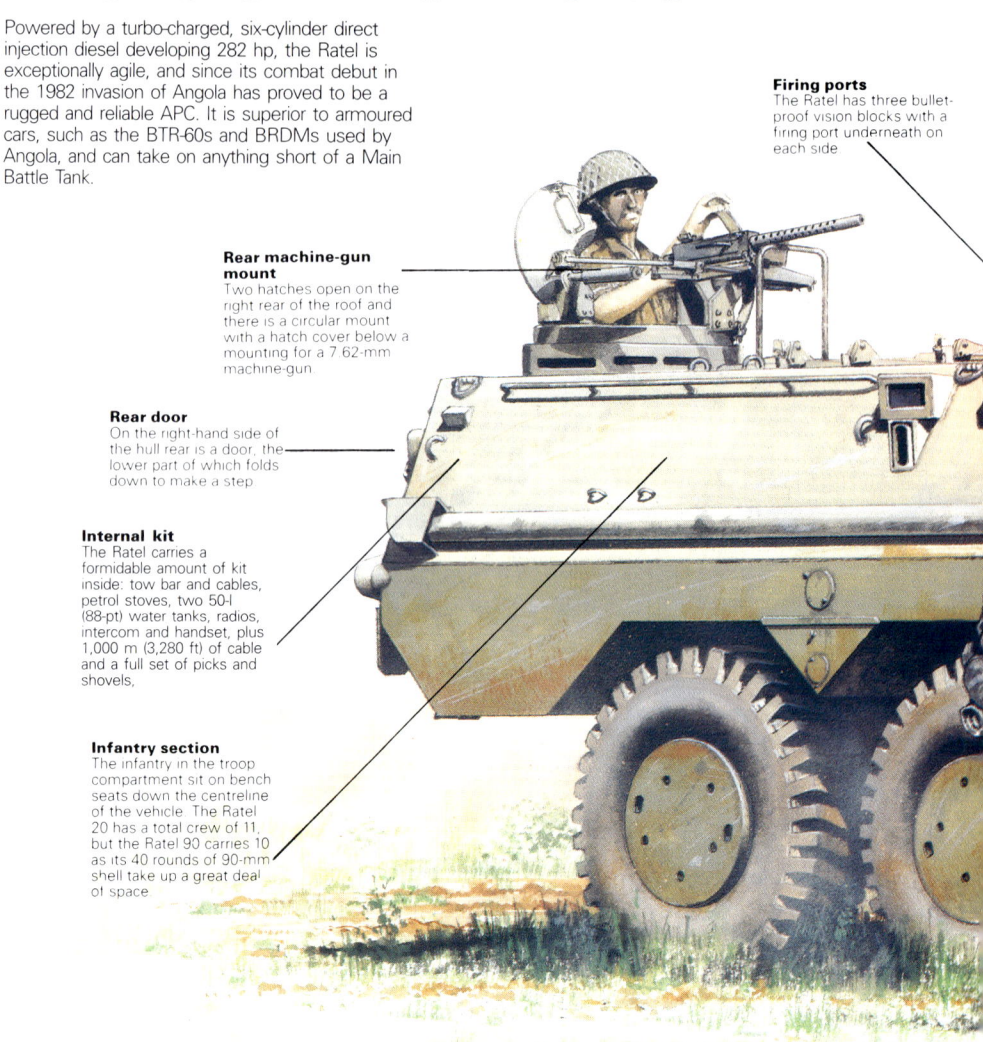

Powered by a turbo-charged, six-cylinder direct injection diesel developing 282 hp, the Ratel is exceptionally agile, and since its combat debut in the 1982 invasion of Angola has proved to be a rugged and reliable APC. It is superior to armoured cars, such as the BTR-60s and BRDMs used by Angola, and can take on anything short of a Main Battle Tank.

Firing ports
The Ratel has three bullet-proof vision blocks with a firing port underneath on each side.

Rear machine-gun mount
Two hatches open on the right rear of the roof and there is a circular mount with a hatch cover below a mounting for a 7.62-mm machine-gun.

Rear door
On the right-hand side of the hull rear is a door, the lower part of which folds down to make a step.

Internal kit
The Ratel carries a formidable amount of kit inside: tow bar and cables, petrol stoves, two 50-l (88-pt) water tanks, radios, intercom and handset, plus 1,000 m (3,280 ft) of cable and a full set of picks and shovels.

Infantry section
The infantry in the troop compartment sit on bench seats down the centreline of the vehicle. The Ratel 20 has a total crew of 11, but the Ratel 90 carries 10 as its 40 rounds of 90-mm shell take up a great deal of space.

Ratels are shown being serviced after returning from an operation. Ratel columns perform a great deal of maintenance in the field, and their already cramped interiors are further crowded by the addition of spare parts and enough supplies for several weeks' action.

Fire support vehicle

There are many different Ratels – the Ratel 90 is a fire support vehicle with a 90-mm turret gun, and the Ratel 60 turret carries a short breech-loading mortar, again for fire support, but it is also a powerful anti-ambush weapon that can fire a canister round to produce a fan of steel balls over a wide area at short ranges.

There is a command version of the Ratel with a 12.7-mm Browning heavy machine-gun in the turret and a Ratel mortar carrier with a 81-mm mortar firing through the roof hatches. It is anticipated that a version carrying some form of anti-armour guided missile will be seen in the near future.

All these various forms of Ratel operate over very long ranges, often deep into neighbouring states around the borders of South Africa to knock out terrorist strongholds and bases;

MECHANISED COMBAT

81-mm smoke dischargers
These can be fired by either the gunner or commander and quickly provide a dense smokescreen to conceal the Ratel from the enemy.

Commander
The commander sits on the left of the turret with the gunner on the right. The commander's cupola has vision blocks for all-round observation while closed down.

Two-man turret
Ratels are fitted with the same turret as carried on the Eland armoured cars used by the South African Army. This example is fitted with a 20-mm cannon and co-axial machine-gun, but the FSV 90 carries a 90-mm semi-automatic gun firing HEAT rounds to an effective range of 1,200 m (3,940 ft).

20-mm F2 cannon
Turret traverse and weapon elevation are manual on the Ratel 20, the cannon can elevate to +38° and depress to -8°. It fires High Explosive rounds to an effective range of 2,000 m (6,560 ft) and armour-piercing ammunition to 1,000 m (3,280 ft).

Roof hatches
There are four roof hatches above the troop compartment which are hinged on the outside and can be locked open to provide welcome ventilation.

Driver's station
The driver has excellent visibility through three big bullet-proof windows. In action, these are covered by steel shutters at the touch of a single lever within the vehicle. The driver's station is connected to the rest of the vehicle so you can change drivers without anyone having to get out.

All-welded steel hull
The Ratel's side armour is 8-10 mm (0.3-0.4 in) thick, proof against small-arms fire up to 7.62 mm calibre and shell splinters. The frontal armour is 20 mm (0.8 in) thick and will keep out 12.7 mm machine-gun rounds, but the 14.5-mm and 23-mm anti-aircraft weapons sometimes encountered defending guerilla camps can destroy a Ratel.

Below: Smoke break for the commander and gunner of a Ratel 20 leading a column of Buffel mine-resistant personnel carriers.

the most frequent targets have been SWAPO bases in Angola.

During such operations the Ratels are loaded with all manner of equipment and supplies. Spare wheels are carried lashed to the hull roof, food is stacked wherever there is space, and there always seems to be room for a few cases of beer. The usual number of 7.62-mm machine-gun rounds carried is at least 6,000. Spare whip aerials for the radio are always carried somewhere on the hull, as these break continually when Ratels shove their way through dense bush.

Equipment fit

Two drinking water tanks are fitted as standard, and each vehicle carries cooking stoves, a comprehensive tool kit, a tow bar and cables, and spare parts. At least one vehicle in every four-Ratel troop tries to carry a field shower outfit. The overall emphasis is on self-sufficiency from fire support to first aid, for out in the border country there are no nearby bases and no supply dumps, other than those which can be captured.

Self-sufficiency

The self-sufficiency is carried to the point of changing engine packs in the field. Each Ratel column is usually followed by a convoy of trucks, some of which carry repair crews who can replace a Ratel engine pack in about 30 minutes. The convoy also carries fuel and ammunition and some other supplies, but, in the main, each Ratel has to carry its own needs for sometimes well over two weeks.

Consequently Ratel interiors are crowded and cramped while on

MECHANISED COMBAT

operations, and it comes as no surprise that the roof hatches are kept open as long as possible with the crew spending most of their time sitting on the roof or with their heads out of the hatches.

In action, Ratel crews can either fight their way through an objective by firing their weapons through the weapon ports as they drive through, using the turret guns to add to their firepower, or they can dismount for action, in which case the turret guns are used as mobile fire support.

Effective firepower

The 20-mm cannon and 90-mm guns carried by most Ratels have a useful anti-armour capability and can knock out most of the lightly-armoured vehicles or strongpoints that they are likely to encounter, and they are also very effective anti-personnel weapons. The Ratel 60s and mortar-carriers tend to stand back during the final stages of attack operations to supply indirect fire support, but they too can use their weapon ports to provide more firepower.

Most Ratel operations are fairly simple in military terms. Columns up to a battalion strong leave their base somewhere near a border and make deep incursions into neighbouring territory. The exact positions and nature of the terrorist bases involved are already well known from the efforts of recce teams who have scouted the locations and approach routes for weeks in advance.

Native scouts

These teams often operate on foot and usually involve local natives who know the country well. Using their guidance the column makes its approach at speed and across country to avoid land mines on the roads and tracks. As it gets nearer the objective, most moves are made by night.

Guns blazing

The final approach is made under cover of darkness with the attack commencing at dawn by all available vehicles charging forward in line abreast with all guns blazing. The actual fire fights usually last only minutes, with the guerrillas fading into the local country to hide and come back another day – they are short of anti-tank weapons and can do little against the well-armed Ratels and their well-trained crews.

Not all Ratel work is of such a dramatic nature. Many Ratels are used for little else than escorting convoys and border patrolling to keep an eye on the vast tracts of open country that guerrillas could use to infiltrate peaceful areas. During these patrols each Ratel has an operational range of about 1,000 km (620 miles), so they rarely need refuelling as would similar vehicles.

Robust vehicle

One thing marks the Ratel above all others – its extreme toughness. It can operate under conditions that would wreck many similar vehicles, and it can travel great distances without breaking down. Yet when the Ratel enters a fight it can hold its own against any comers.

The Ratel has a 'V'-shaped inner hull to give its crew and passengers some protection against mines. Most Ratels are armed with 20-mm cannon, but the lead Ratel here is being used as transport and has no gun.

MECHANISED COMBAT

RIDING INTO BATTLE

Modern armoured forces are a combined team, with tanks and infantry fighting in close co-operation. The infantrymen ride into action in Armoured Personnel Carriers (APC) which can keep pace with the tanks; sometimes you will fight from the vehicle, but in most situations you dismount to fight on foot. Certainly, the mechanised infantry platoon with its four APCs has far greater freedom of movement and can respond far quicker than 'leg' infantry.

In attack, the leader will try to fight from the vehicles for as long as possible, using the tactics worked out in advance for just this sort of situation, and will only get his men out of the protection of the APCs when he gets into close terrain like trees and bushes, or comes up against obstacles or a strong anti-tank force.

This flexibility – to fight from vehicles with armour strong enough to deflect small-arms fire or to dismount and take on anti-armour forces with conventional infantry tactics, then perhaps call up the APCs, mount up again and carry on the advance as before – has meant that a new way of infantry fighting has had to be developed.

Make your position secure

A typical APC is 2.5 m (8.2 ft) tall and 5 m (16 ft) long, so you choose lying-up and ambush positions with a

An Armoured Personnel Carrier enables you to move swiftly, keeping up with tanks. These US M113s carry the increased machine-gun armament adopted during the Vietnam War: a .50-cal Browning at the front, and an M60 7.62-mm machine-gun on each side.

6 BASIC RULES OF MOVEMENT

1. Make use of terrain that hides you from enemy observation or fire.
2. Avoid silhouetting your vehicle by crossing a skyline or moving directly forward from a hull-down position.
3. Cross open areas of ground as fast as you can.
4. Use your smoke grenade-launchers to cover disengagement or to protect a halted APC.
5. Move with a small force scouting ahead and with the rest of your team following behind.
6. Make sure your leading team can be covered by the vehicles behind.

MECHANISED COMBAT

lot of thought about how you're going to conceal it, both by means of natural cover and by applying camouflage.

Hollows in the ground, buildings and courtyards and patches of mature undergrowth and young trees (which are vulnerable to the vehicle itself if you have to take off in a hurry) are all likely locations.

Because you can't 'stand-to' an APC in just a second or two as you can an infantry fighting unit, it's especially important to get your local security well established. Reconnoitre the area very carefully before committing yourself to a location, and keep the patrol activity up all the time the vehicles are in position.

Static sentry positions should be further out than normal, and this may make for a communications problem.

Remember, when selecting sites for individual vehicles, that you're not just parking them. They should be able to support and defend each other and the dismounted troops that are put out to defend them. For mounted infantry to be really effective, the men and the machines have to be working together as a team.

Move fast

Just like any infantry formation, APC-mounted infantry rely a lot on the principle of fire and movement – the squad divides into two; one part puts down fire while the other moves, then they change jobs, and so on.

But there are two main differences when you're using an APC – its machine-gun or cannon can operate at longer range, and the movement takes place at 10 times the speed. The squad leader – and the team leaders, who now become vehicle commanders as well – must make sure that these features of the APC are used to the full, and this means careful planning and preparation.

Just like tanks – and also anti-tank helicopters – APCs are best sited in 'hull-down' positions. All their armament is found on top of the vehicle, while the infantry riding in them uses a door at the rear to mount and dismount.

Obviously, if the troops can get in and out while protected from enemy fire, the whole operation will be a great deal safer. If the vehicles can move in the hull-down position as well – along roads or tracks with hedges and banks on each side, for example – then they are very difficult to detect, even when moving. This adds very considerably to their effectiveness, but gives them less room to deploy in case of attack.

Unit commanders must consider all these points when using the APC in attack. The extra speed of the vehicle gives you every chance of over-run-

PLATOON FORMATIONS

Column formation

This is the most frequently used formation, it is the best for road marches, movement in limited visibility or when passing through woods and defiles. You can deploy quickly into other formations and it is the easiest to control.

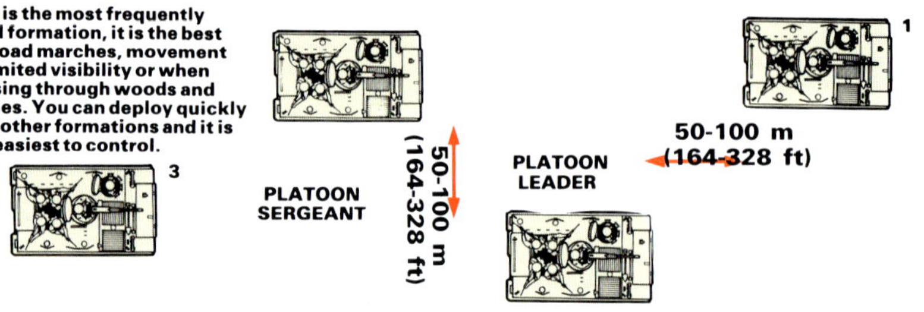

50-100 m (164-328 ft)

PLATOON LEADER
PLATOON SERGEANT

Signals for turning

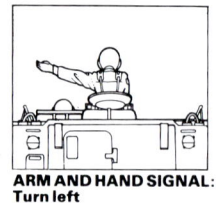

ARM AND HAND SIGNAL: Turn left

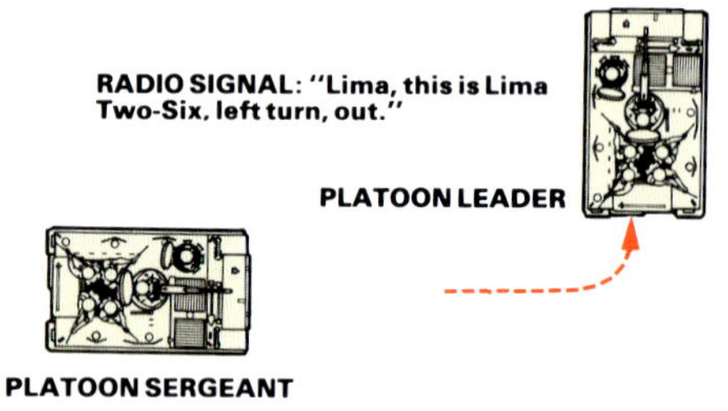

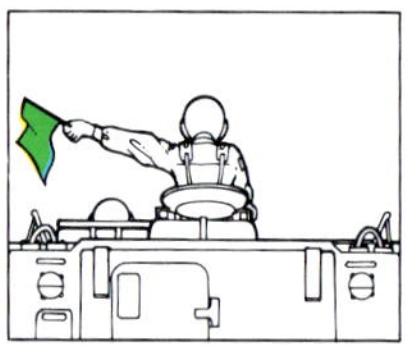

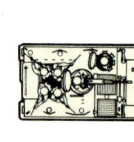

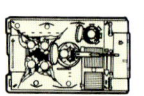

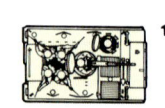

FLAG SIGNAL: Turn left (The flag is green on one side and yellow on the other: green = turn left, yellow = turn right)

RADIO SIGNAL: "Lima, this is Lima Two-Six, left turn, out."

PLATOON LEADER
PLATOON SERGEANT

Line formation

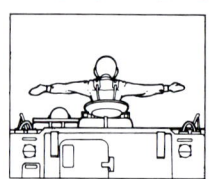

ARM AND HAND SIGNAL

FLAG SIGNAL

This is used when assaulting an objective, crossing open areas, exiting a wood or when emerging through a smokescreen. It gives maximum firepower to the front and is the best way in which to rapidly cross an open area.

Echelon formation

ARM AND HAND SIGNAL

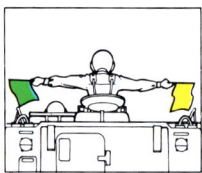

FLAG SIGNAL

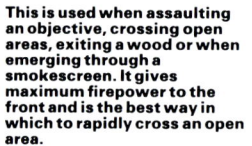

Use echelon when you are on an exposed flank. It gives you excellent firepower to both the front and both flanks.

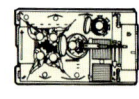

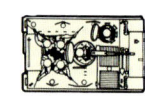

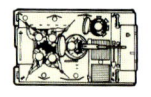

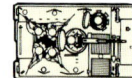

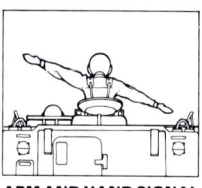

50-100 m (164-328 ft)

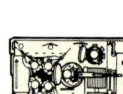

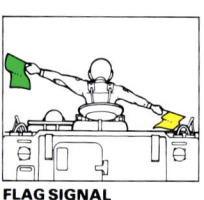

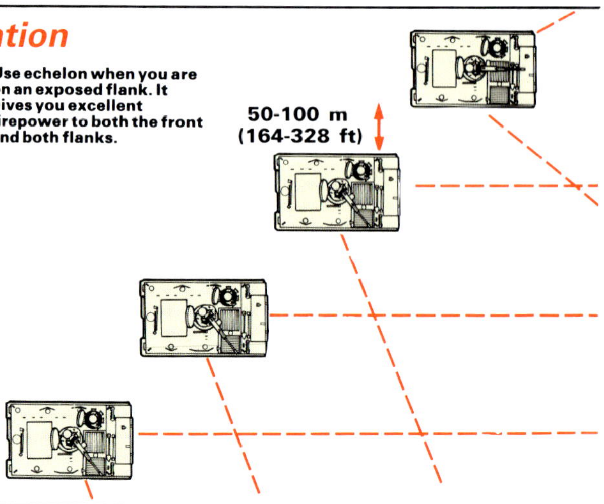

MECHANISED COMBAT

Vee formation

Use this when the situation is unclear and you want the unit concentrated with all-round firepower.

ARM AND HAND SIGNAL

FLAG SIGNAL

Wedge formation

This is easy to control and is simply a line with the flanking vehicles echeloned back. It is also used when the situation is unclear and you may need to deploy to either flank.

ARM AND HAND SIGNAL

FLAG SIGNAL

ALTERNATE SIGNAL

PLATOON LEADER

50-100 m (164-328 ft)

50-100 m (164-328 ft)

PLATOON SERGEANT

Herringbone formation

This is adopted by a column when it needs to deploy quickly, e.g. if ambushed or facing an enemy air attack. It proved very effective in Vietnam.

ARM AND HAND SIGNAL

PLATOON SERGEANT

PLATOON LEADER

Coil formation

Coil is a stationary formation providing all-round defence. It is used for refuelling, re-supply and giving orders. It should not be used for very long in daylight as it presents a concentrated target.

ARM AND HAND SIGNAL

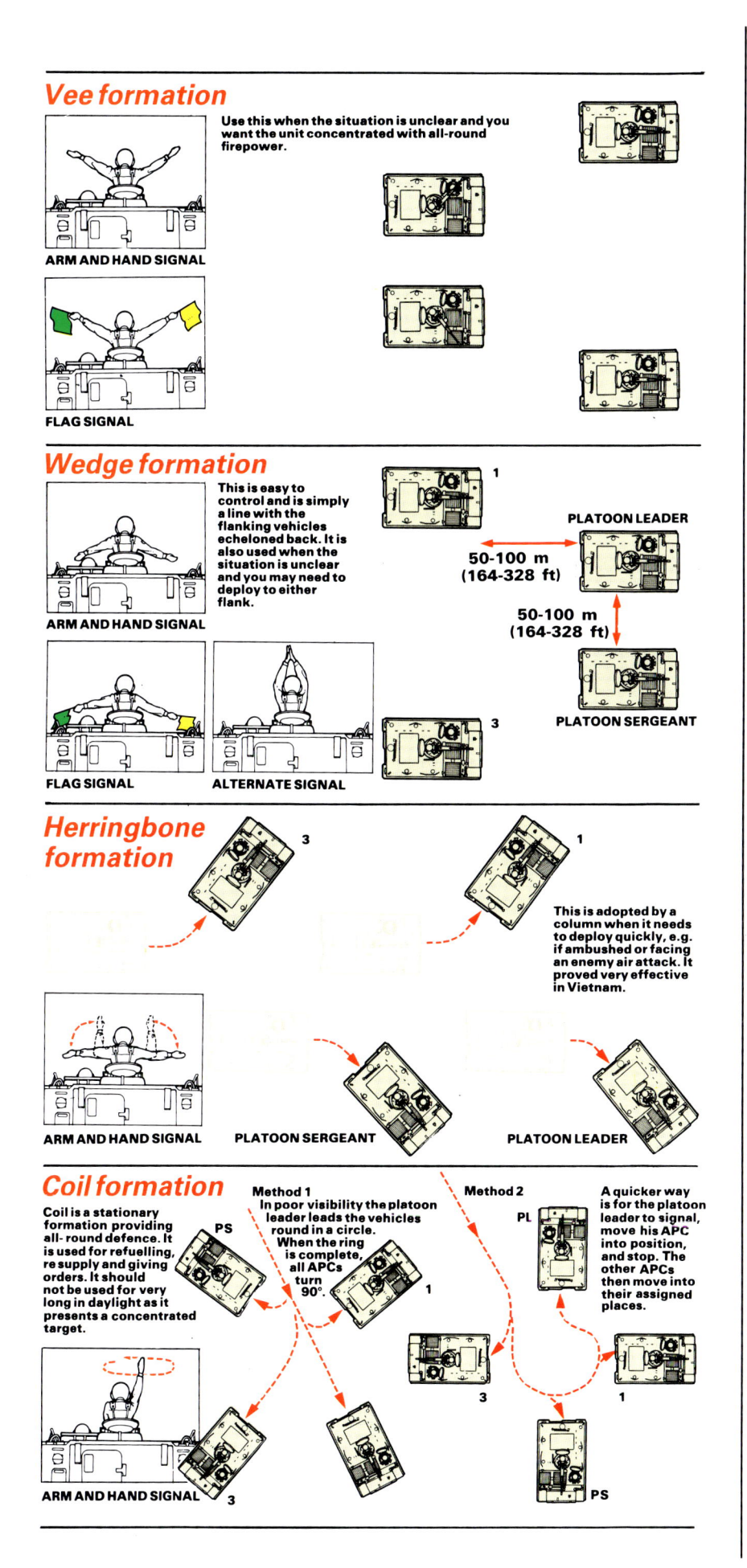

Method 1
In poor visibility the platoon leader leads the vehicles round in a circle. When the ring is complete, all APCs turn 90°.

Method 2
A quicker way is for the platoon leader to signal, move his APC into position, and stop. The other APCs then move into their assigned places.

If an M113 strikes an anti-tank mine anyone inside is likely to be injured. These Australian troops are riding on top where they are safer from mines, but of course more vulnerable to enemy rifles.

ning enemy positions – especially if they've been careless in their anti-armour preparations – but it also means that it's easy to over-extend, to get so far in front that the advance becomes a series of isolated fire-fights that do little or nothing to really gain ground, and where you're in every danger of being surrounded and cut off.

Keep in touch

Communications between vehicles often require a radio net. As well as the sets fixed into the APC, the platoon commander, the platoon sergeant and each team-leader will have personal radio transceivers. This means that communications are usually better between members of a mounted infantry unit than between foot soldiers in a squad, again making for better mobility and quicker response times.

It does make for one added danger, however – the enemy may be able to listen-in to your transmissions. If he does he will not only gain intelligence, but also be able to pinpoint your position.

Sustaining the attack

Because APCs are at risk from even hand-held anti-tank weapons, it's most important to allow the enemy no time to re-group and get its anti-armour specialists into the fire-fight. The speed at which the M113 can move cross-country gives the mounted infantry unit commander an advantage here, but he is still just as concerned to keep a high rate of fire concentrated on enemy positions. Getting from place to place quickly is important, but it's still weight of fire that wins fire-fights.

He has to think about re-supply, too – ammunition, food and one new factor: fuel for the vehicles. Get too far

MECHANISED COMBAT

away from a supply point, and you could suddenly find yourself helpless, with your carriers out of fuel. At that point, all the advantages you've had suddenly turn into liabilities.

Deadly missiles

The APC's worst enemy is the Anti-Tank Guided Missile (ATGM), now so light and compact that you must expect even small units of enemy troops to be equipped with them. Missiles such as these have one big weakness: they don't work well if there are obstacles – trees, for example, or even wire fences – between the launcher and the target. In open country, though, they're deadly.

It is the APC's driver who is the vehicle's first line of defence against ATGMs. His skill at using the shape of the country to keep the vehicle out of the sight-line of enemy troops, and his ability to keep the vehicle moving through difficult patches instead of cutting across open country, make all the difference.

Terrain driving, as it is called, is practised over and over until it becomes second nature following four very basic guidelines:

1 Use all available cover.
2 Avoid the skyline.
3 Cross even small open areas fast.
4 Don't move straight forward out of a hull-down firing position.

Even though all but the last of these are basic skills that every infantryman

A combined arms team of M48 tanks and M113 APCs halts at the edge of a forest and the dismount teams prepare to assault on foot. In Vietnam armoured forces proved more effective than expected in jungle.

learns, the way they're put into practice is changed a lot by the size and speed of the vehicles. Reading the terrain, whether from the map or from looking directly at the ground, becomes even more important.

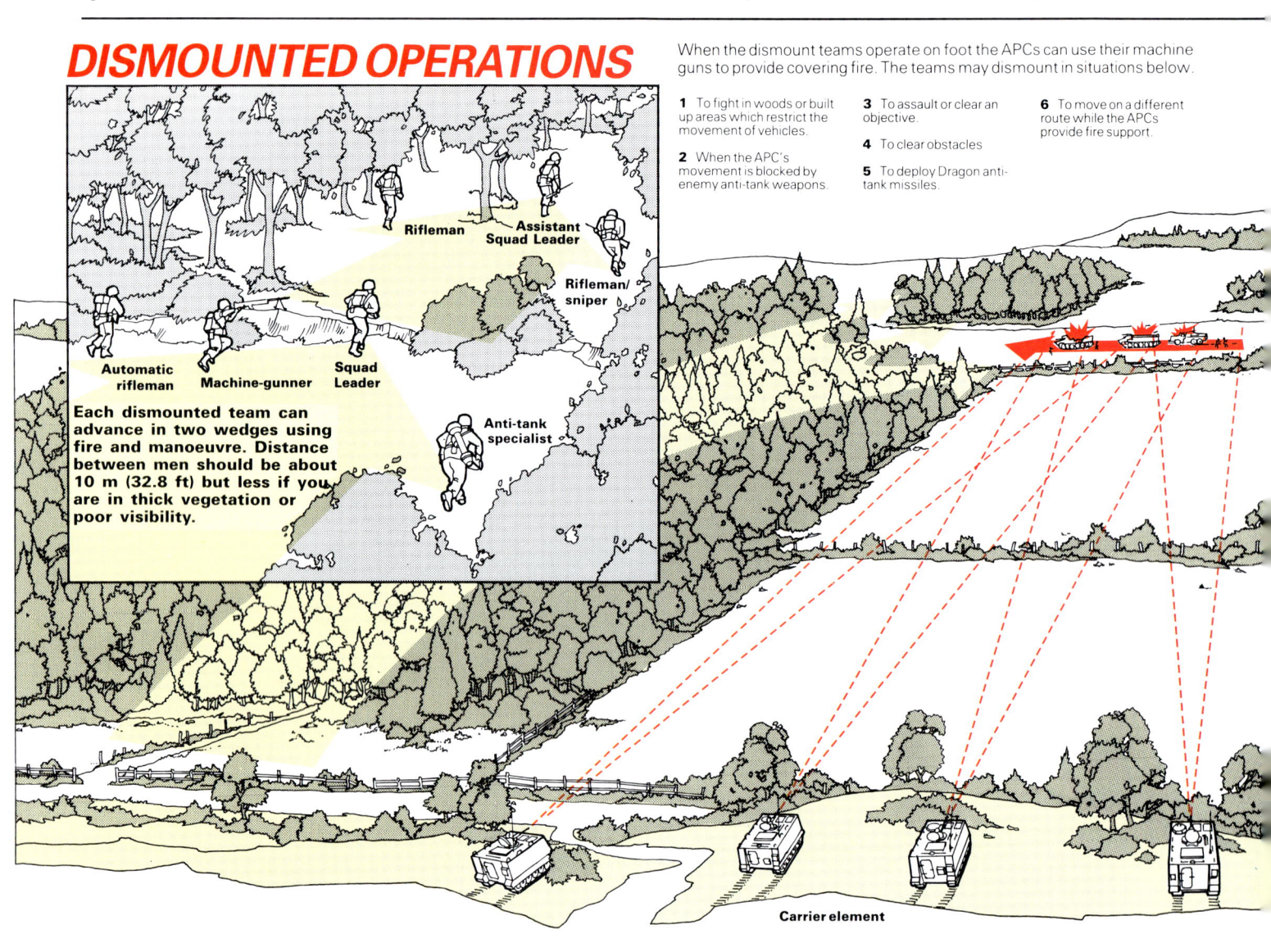

DISMOUNTED OPERATIONS

When the dismount teams operate on foot the APCs can use their machine guns to provide covering fire. The teams may dismount in situations below.

1 To fight in woods or built up areas which restrict the movement of vehicles.

2 When the APC's movement is blocked by enemy anti-tank weapons.

3 To assault or clear an objective.

4 To clear obstacles

5 To deploy Dragon anti-tank missiles.

6 To move on a different route while the APCs provide fire support.

Each dismounted team can advance in two wedges using fire and manoeuvre. Distance between men should be about 10 m (32.8 ft) but less if you are in thick vegetation or poor visibility.

Rifleman, Assistant Squad Leader, Rifleman/sniper, Automatic rifleman, Machine-gunner, Squad Leader, Anti-tank specialist

Carrier element

MECHANISED COMBAT

CASCAVEL
Armoured Car from Brazil

The ENGESA EE-9 Cascavel, seen here coming ashore from a Brazilian amphibious assault ship, is one of the most successful products of the fast-growing Brazilian arms industry.

Fast, powerful and readily available, the Cascavel armoured car is being sold all over the world by the flourishing Brazilian arms company, ENGESA. Brazil has cheerfully exported AFVs and rocket artillery to both sides in the Iran-Iraq War, as well as to armies in Africa and Latin America. The first Cascavels appeared in 1972, produced for the Brazilian army. Since then some 2,500 have been manufactured and it is ENGESA's greatest success.

Armed with a 90-mm gun, the Cascavel is protected by an unusual dual-hardness armour configuration consisting of an outer layer of hard steel and inner layer of roll-bonded, heat-treated softer steel to afford maximum ballistic support. Increased protection is given to the frontal arc, and special attention has been paid to the threat posed by booby traps, grenades and Molotov cocktails, any of which could prove lethal to a crew operating in the confines of a jungle or, more probably, in a built-up area during a period of civil unrest.

The driver, who is provided with a single hatch cover opening to the right, sits in the front left of the hull. Everything possible has been done to make his life comfortable – his seat and steering wheel are adjustable, and there is a small windscreen and wiper that fold forward onto the glacis plate when not in use – but the three periscopes in the top part of the glacis plate provide him with no more than 120° visibility when closed down,

MECHANISED COMBAT

The Boomerang walking beam suspension, in which the rear wheels are capable of extreme vertical travel, gives the EE-9 excellent performance over rough ground. The same system is used on ENGESA's range of 6×6 tactical trucks.

barely enough for operational purposes.

The Boomerang walking beam suspension, originally fitted to ENGESA 6×6 trucks, is versatile and efficient, consisting of a rigid axle connected to the hull by double leaf springs and telescopic dampers holding two lateral walking beams through which power is taken from the drive to the four rear wheels. Up to 0.9 m (3 ft) of vertical travel is built into the rear wheels, enabling the vehicle to keep them in contact with the ground at all times, which considerably enhances the vehicle's traction when operating in muddy, hilly conditions prevalent in parts of South America. As a refinement, the run-flat tyres will keep the Cascavel mobile for over 100 km (62.5 miles) even when fully deflated.

Optional equipment, including the air-conditioning system, heater, laser rangefinder, automatic fire extinguisher and active or passive night vision enhancements, are available and, when fitted, upgrade the Cascavel considerably, bringing it up to European and United States' standards.

Powerpack

A Detroit Diesel 6V-53N six-cylinder water-cooled diesel engine developing 212 hp at 2,800 rpm is normally fitted, although alternatives may be installed if required. The versatile and powerful Detroit engine is capable of a maximum road speed of 100 km/h (62.5 mph), but is economical enough to give a maximum range of 880 km (547 miles), far greater than the majority of more sophisticated vehicles. Although not amphibious, the Cascavel can ford to a depth of 1 m (3.3 ft), can climb gradients of 60 per cent and overcome vertical obstacles of 0.6 m (2 ft). The engine, at the rear of the hull, is co-located with the Detroit Diesel MT 643 gearbox, which has four forward and one reverse gears. A purpose-built ENGESA transfer box splits the drive between the front and rear differentials, giving maximum traction and control whatever the speed and conditions.

Of prime importance when operating in adverse conditions, access to the engine is easily obtained via two large, hinged doors at the top rear of the hull. Although the engine cannot be removed as quickly and easily as in the case of many armoured vehicles of European construction, particularly the French AML or British CVR(T) series, this is not in itself a problem in South American armies, which cannot always rely on advanced mechanical assistance in the field. The engine design is basic enough and access adequate to enable most immediate problems to be resolved.

A double 24-volt electrical system is fitted, one part of which generates power for the engine and turret. The other acts as a fail-safe back-up to start the engine after long periods of static radio operations — again a concept sadly lacking in many more sophisticated and far more expensive designs.

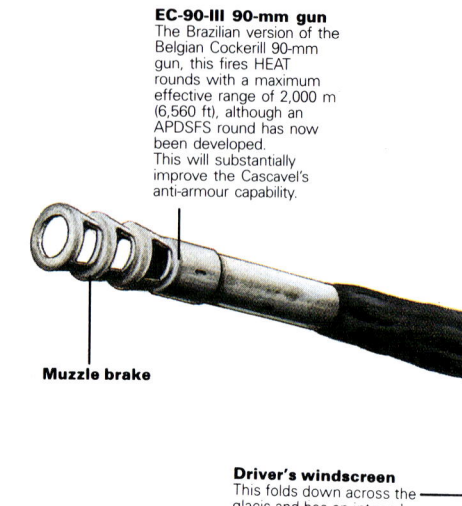

Firepower

Initial production models of the Cascavel were fitted with a French H-90 turret armed with a 90-mm gun similar to that equipping the highly successful AML 90. Current models, however, boast the domestically produced ENGESA ET-90 turret and EC-90 gun. Designed with the export market in mind, both have been installed successfully in the Greek Steyr 4K 7FA APC, have been trialled with the Vickers Valkyr and will fit a wide

The commander's cupola stands 2.6 m (8.5 ft) off the ground, making an imposing sight, but a potentially good target. The Browning .50 cal, mounted by the cupola, is fitted in a similar manner to the Russian 12.7 mm AA machine-guns. Ground clearance is 34 cm (13.4 in) to the front axle and 50 cm (19.7 in) to the hull.

EC-90-III 90-mm gun
The Brazilian version of the Belgian Cockerill 90-mm gun, this fires HEAT rounds with a maximum effective range of 2,000 m (6,560 ft), although an APDSFS round has now been developed. This will substantially improve the Cascavel's anti-armour capability.

Muzzle brake

Driver's windscreen
This folds down across the glacis and has an integral windscreen wiper.

Dual-hardness armour
Protected over the frontal area by up to 16 mm (0.63 in) of steel armour, the Cascavel is proof against small-arms fire and machine-guns.

Combat tyres
The tyres are run-flats and enable the Cascavel to travel up to 100 km (62.5 miles) on deflated tyres.

MECHANISED COMBAT

Inside the Cascavel

The Cascavel was designed by ENGESA to meet the requirements of the Brazilian army. First production vehicles were armed with 37-mm guns removed from the ancient M3 Stuart light tanks acquired in the late 1940s. The next batch sported French H-90 turrets, but current production models have a Brazilian turret and the vehicle has been widely exported.

Browning .50-cal machine-gun
Shown here on the basic anti-aircraft mounting, other fittings are available, including one that allows the gun to be fired from inside the turret.

ser rangefinder
e first Cascavels to carry ser rangefinder carried ere above the barrel, this mounting is erable to shrapnel and all arms fire and they now built into the ner's sight.

Commander's cupola

Commander

Smoke grenade launchers

Central tyre pressure regulator
The driver can adjust tyre pressure to suit the type of ground the vehicle is traversing. On roads tyre pressure should be 4 kg/cm² (57 PSI); going cross-country, 3 kg/cm² (42 PSI); and 2 kg/cm² (28 PSI) on snow, mud or soft sand.

Boomerang suspension
With 90 cm (35.4 in) vertical wheel travel, the walking beam suspension enables all four rear wheels to stay in contact with the ground, however rough the terrain. This contributes enormously to the Cascavel's cross-country performance.

series of other tracked or wheeled armoured cars and personnel carriers. The all-welded steel turret, 16 mm (0.63 in) thick at the front and 8.5 mm (0.3 in) thick at the sides, rear and top, is proof against all small-arms fire and anything other than a direct hit from 105-mm light artillery. The gun is able to penetrate protective armour of any APC and of most reconnaissance vehicles. This makes the Cascavel a formidable fighting machine in any Third World conflict, in which it is likely that expensive and sophisticated modern Main Battle Tanks will not be deployed.

The turret itself is comparatively low but is surmounted by a huge cupola, presumably inspired by the French AMX design, above the com-

The Cascavel has a laser range-finder built into the gunner's sight (the hooded screen next to the cupola) instead of mounted over the gun. It is protected by a shutter which falls into place when you fire the main armament.

87

MECHANISED COMBAT

Brazil is a country of wide climatic variation, and vehicles such as the Cascavel are designed for operations in terrain ranging from Amazon jungle to the slopes of the Andes mountains.

mander's seat on the left. Both the turret and cupola consist of an outer layer of hard steel and an inner layer of steel roll-bonded and heat-treated for enhanced ballistic protection.

The main armament consists of the Cockerill-designed, ENGESA-manufactured EC-90 gun with an elevation of +15°, depression of −8° and manual traverse through 360°. A powered system does exist as an expensive alternative.

An externally mounted 7.62-mm machine-gun, aimed and fired from within the safety of the turret, can be mounted onto the ENGESA ET-762 cupola and in certain instances can be replaced by the far more potent 12.7-mm M2 HB machine-gun mounted on a Soviet-style DShKM anti-aircraft mantel.

Due to its comparatively small size, the turret is a mere 1.84 m (6.04 ft) wide and 0.59 m (1.93 ft) high. Storage space is at a premium, reducing the weapon-load to a rather unsatisfactory 24 rounds of main ammunition and 2,000 rounds of 7.62-mm for the machine-guns. This means frequent replenishment where possible or, more likely, strict ammunition conservation, a discipline rarely found among Third World armies.

Although a Mk V model of the Cascavel retro-fitted with a Mercedes-Benz OM 352 A diesel engine developing 190 hp at 2,800 rpm is available, no true variants exist. But a number of optional extras are available, including a recently developed laser rangefinder aimed through the gunner's sight and protected by a shutter when the main armament is fired to take the place of the externally mounted and therefore vulnerable external sight, and various radio installations.

Basic equipment

The Cascavel does not have any form of NBC protection – presumably not a problem in the environments in which it is likely to operate – nor does it have the benefit of additional air-conditioning, which would be of distinct benefit to crews operating in the uncompromising heat of the North African desert.

To date, the EE-9 Cascavel has proved a great success. It shares many of its automotive parts with the EE-11 Urutu armoured personnel carrier, making a combined purchase financially attractive, and can be fitted with a smaller 300-mm recoil 90-mm gun and turret if required. By January 1984, 2,550 EE-9 Cascavel armoured cars had been produced and construction continues unabated, although current production figures have not been revealed. While the major powers continue to produce complex combat vehicles beyond the pockets of smaller nations, weapons systems such as the Cascavel will always have an assured market.

ENGESA manufactures the EE-11 Urutu APC, which shares many features with the Cascavel. Both vehicles incorporate commercially-available automotive parts, which greatly reduce manufacturing and operating costs.

The EE-11 has the same dual-hardness armour, Boomerang suspension system and run-flat tyres as the EE-9. It is available with a wider variety of armament, however, and can be equipped with anti-tank missiles.

MECHANISED COMBAT

Camouflaging Your Vehicle

Good camouflage and concealment is often a trade-off against good fields of fire or good positions for observing enemy movement. Radio communications work better with line-of-sight, but sitting on top of a hill is very public. And if you are trying to evade or escape you will need a vantage point for a sentry to observe likely enemy approaches, and may be observed yourself.

Assuming that you are part of a group of six to 12 men and that you have a light vehicle like a Land Rover or jeep, how would you conceal your position while evading capture?

Siting

Avoid the obvious. If the enemy are looking for you they will sweep the countryside, and if there are not many of them they will concentrate on rivers and woodland, farm houses, barns, known caves and natural cover. All are on maps, and the first move that an enemy search team will make is to do a map reconnaissance and look at likely locations.

Track plan

A track plan is essential if you are going to stay in the location for any length of time. Trodden grass and

A little camouflage can go a long way; these vehicles look like patches of gorse at a distance. Concealment on the move is dependent on using cover and dead ground.

Selecting a vehicle hide

1 Site selection
Left: Choose a harbour area away from the edge of the wood, away from tracks and with good cover overhead as well as at ground level. Try to pick a 'hull down' or 'dead ground' position. Remember to back the vehicle in; you may have to exit fast.

2 Hessian sacking
Above: All the principles of personal camouflage apply equally to your vehicle. Black hessian destroys the shine from windows, headlights and number plates, and disguises the general shape of the vehicle.

MECHANISED COMBAT

footprints will show clearly from the air, and large areas of normally lush undergrowth can be flattened in a way that attracts attention.

Vehicle tracks are even more dramatic from the air – bad drivers will carve a path across a field in a way that no farmer would dream of driving. Track planning means attempting to copy the normal routes adopted by animals, farmers or the locals. Thus vehicle tracks along the edge of a field and a footpath that might also be used by the inhabitants will pass unnoticed by the enemy.

IR signature

As with personal camouflage, the infra-red band is the most difficult to avoid. Thermal imaging will penetrate cover, and activities like running a vehicle engine to charge batteries or simple tasks like cooking become a major problem since both will show as a very strong hot point in an otherwise cool terrain.

Though a cave may not be ideal if it is on the local map, it will give good thermal screening. Parking the vehicle under cover will also reduce its IR signature – but again remember that barns and farmhouses are very obvious and may attract attention from the air or ground.

Concealing your vehicle

Any vehicle will be under suspicion. If you are moving in convoy, take care to avoid bunching. Vehicles close together are very recognisable from the air, and make easy targets for enemy aircraft. And remember the following points when finding somewhere to position your vehicle.

1 If you are near buildings, for instance on a farm, try to get the vehicle close to a wall or under cover in a barn. A camouflage net will attract the attention of a nearby enemy; use hessian and local materials to disguise the vehicle.

2 If you park in the country, try to find the shadow of a hedge to disguise the vehicle's hard shape. But remember that in northern and southern latitudes the sun moves, and the shadow of the morning can be the sunlit field of the afternoon.

3 Late evening can be particularly difficult, with low sunlight catching the glass fittings of your vehicle. As a short-term precaution, cover the windscreen and lights when you stop, not forgetting the reflectors.

4 If your vehicle is military, it will have been painted with IR reflective paint and you should not cover this with hessian, which will produce a blue-grey colour on any infra-red device that the enemy might be using. You should cover the reflective surfaces and then deploy a camouflage net.

5 A camouflage net should stand clear of the vehicle, partly so that you can get in and out and also to disguise the vehicle's shape. It should also stretch far enough to contain any shadow that the vehicle might cast. Ideally, it should also have a 'mushroom' on the top: a frame of wire about the size of a domestic saucer. This gives a smooth line when the net is stretched over. Make sure that the net will not snag on the vehicle or underbrush or trees, preventing any quick exit you might need to make.

4 Camouflage net
Above: Use the surrounding trees as well as the poles. The ideal situation is to create a camouflage 'garage' you can drive into and out of without having to remove net, poles etc.

3 Net poles
Above: A good selection of net poles is essential to hold the camouflage net off the vehicle to disguise its shape. Chicken wire can also be used. You must not cut poles from trees around your position; the cut-off shoots will give you away. Harbouring two vehicles together with nets over both can be helpful in producing a more natural shape. Remember, you cannot afford to leave any equipment lying about; concealment is an ongoing task, as the threat of discovery is ever-present. Plastic bags and uncovered windscreens are asking for trouble.

Sound and smell

As with personal camouflage, sound and smell are important. If you run your engine to recharge batteries, you will make noise and exhaust fumes (and take care that fumes do not blow into the vehicle if the exhaust pipe is blocked by the camouflage). Use a flexible metal extension pipe to reduce the noise.

If you are in a convoy, the sound of your vehicles will attract attention, and so will your radio traffic.

Smell will come from cooking as you prepare your evening meal, and the smell of fuel is also distinctive. Spilled fuel and the wrappings from rations are a calling-card for an alert enemy.

MECHANISED COMBAT

Concealing your position

Don't make the mistake of thinking you're safe as long as you have dug your position. A good hide or bunker should be invisible even at close quarters; if you have dug it well and are careful in your movements, it may pass unnoticed. But the enemy can still spot you if you haven't been careful enough: keep the following in mind.

1 The colour of soil that has been dug from lower than about a metre is lighter than the topsoil, and a trench has a strong shadow at the bottom. Conceal earth by covering it with turfs; and put light-coloured straw at the bottom of a trench to reduce some of the shadow. This will also be more pleasant to walk on and live in.
2 In a tropical environment, cover can grow very quickly, so replace plants and creepers around your position and it will soon be concealed.
3 A simple basha made up with poncho or basha sheet can be square, shiny and noisy. Do not put it up until after last night – although you can position it flat on the ground before dark. Carry a length of old camouflage net; it will break up the shape and shine.
4 When you are cooking or brewing up, keep your opened kit to a minimum; you might need to make a quick getaway. Also, avoid littering tins and wrappers around the position that may catch the light and be seen from a distance.
5 It is commonly thought that a hand torch with a red filter does not show at night. It does; it's certainly less obvious than a white light, and it does not impair night vision, but it shows. Do not use a torch at all; by last light you should have set up your position so that your kit is packed and you can reach for your weapon, webbing and pack without needing one.

5 Two-sided net
Above: There are two sides to a camouflage net, with different colour combinations, so use the side that best matches your surroundings.

6 Shell scrapes and track plan
Right: As soon as the position is occupied, a route around the site must be marked by cord and cleared. By using this trackplan, disturbance of natural ground is minimised. Shell scrapes must be dug in 'stand to' positions.

Association is also important – radio antennas around a position or on a vehicle show that it is of significance. Antennas can also catch the light and show up as long, hard shadows in an otherwise concealed position. Most antennas can be situated away from the set, so put them on a reverse slope where they are not only invisible to the enemy, but also have some of their signal screened. Failing that, locate them against a building or tree.

Camouflage is a complex and sometimes contradictory skill. There is a reduced TI signature in a building under cover; but buildings attract attention. Hessian should be used on a vehicle among cold buildings; but not in warmer woodland. If you want to remember one rule to camouflage, it is that you should not give the enemy the signal that will make him look twice.

7 Thermal Imaging (TI)
Above and above left: A short-wheelbase Land Rover on the move, with heat radiating from the engine, transmission and wheel hubs. Each type of vehicle has its own TI signature. Note that the image does not go away when the vehicle's engine is switched off; only when it is cold.

MECHANISED COMBAT

How not to do it

Spot the mistakes in this picture; all of them invite an enemy attack.

1. The background deciduous trees are too open and light; it would be better to move right a few metres to the coniferous trees.
2. The position is far too close to the track.
3. Although the use of hessian is good, the absence of net poles means that the enemy will see a Land Rover and tent covered in a net.
4. Where is this man's rifle, helmet and webbing? He will be vulnerable in the event of a sudden attack.

MECHANISED COMBAT

MARAUDING WITH THE MARDER

The Bundeswehr was the first NATO army to introduce an MICV. Marder entered service 12 years ahead of the American M2 Bradley and 17 years before the British Army acquired Warrior. The well-sloped glacis will stop a 20-mm cannon shell.

When the Marder entered service with the West German army in 1971 it represented not only a breakthrough in military technology but also an enormous improvement in the capabilities of the Bundeswehr. The Marder was the first Mechanised Infantry Combat Vehicle (MICV) to enter NATO service. Unlike the contemporary NATO APCs which were designed simply to ferry troops into action, the Marder enabled West German infantry to fight supported by the heavy armament of their vehicle.

Development

Marder took nearly 15 years to develop. In the late 1950s a chassis was developed which could be utilised for a number of basic vehicles including the Jagdpanzer Kanone and Jagdpanzer Rakete tank destroyers, a light reconnaissance tank and an MICV.

Priority was given to the Jagdpanzer Kanone which entered production in 1965 and then to the Jagdpanzer Rakete so that construction of the MICV was delayed until 1967, and the reconnaissance tank was eventually abandoned.

Troop trials for the MICV ran from October 1968 to April of the following year, after which the vehicle was formally adopted and named Marder.

Marder is not amphibious but can ford up to 1.5 m (4.9 ft) without preparation. Here it uses its deep wading kit, distinguished by the schnorkel to the right of the turret. This copes with a depth of 2.5 m (8.2 ft).

93

MECHANISED COMBAT

Marder has a unique sting in the tail: the box above the hull rear houses a remotely controlled MG3 7.62-mm machine-gun. This traverses 180 degrees to cover the whole rear arc and can elevate to +60 degrees.

Production lines were established at Kassel and MaK of Kiel, and an initial order for 2,801 vehicles was placed. However by the time that production was completed in 1975 this number had been increased to 3,111.

The all-welded hull of the Marder provides the crew of four and six passengers with protection from small arms fire and shell splinters, with the front of the vehicle affording complete protection against up to 20-mm projectiles.

Driver's position

The driver, seated at the front left of the hull, has a single-piece hatch-cover opening to the right and is equipped with three periscopes, the centre of which can be replaced by a passive night driving device for operating closed down. An infantryman, usually the section commander, equipped with a single hatch-cover opening to the right but in this instance supported by a single periscope capable of 360-degree traverse,

A top view of Marder reveals the three roof periscopes which allow everyone in the troop compartment to see out. The commander's hatch on the right-hand side of the turret is clearly visible.

MIRA sight

MILAN anti-tank missile

Rheinmetall 20-mm Rh202 cannon
Mounted above the turret like this avoids filling the vehicle with fumes when firing the cannon and it also allows the gun to depress by 17 degrees, which is useful in hull-down positions.

Steel turret
The front of the turret is armoured to withstand 20-mm cannon shells.

Troop compartment
On most Marders this accommodates six infantrymen, but the A1 version has four crew and only five infantrymen. All Marders have an NBC system fitted as standard.

Idler

Dual-tyred road wheels

Diehl tracks with replaceable rubber pads

Drive sprocket

is seated behind the driver.

The six infantrymen in the troop compartment at the rear are carried in comparative comfort, seated three aside and back-to-back to enable them to fire on the move. Two MOWAG-designed spherical firing ports are built into each side of the troop compartment, as are two circular hatches and three periscopes into the roof.

Mobility

Powered by an MTU MB 833 Ea-500 six-cylinder liquid-cooled diesel engine positioned to the right of the driver, the Marder can develop a useful 600 hp at 2,200 rpm. A Renk four-speed HSWL-194 planetry gear-box and stepless hydrostatic steering unit, transmitting power to the tracks via two final drive assemblies mounted at the front of the hull, combine to give the Marder a top speed of 75 km/h (47 mph) forwards or backwards.

With a maximum road range of 520 km (325 miles), coupled with the ability to climb gradients of 60 per cent and to ford to depths of 2.5 metres (8 ft), the Marder has excellent mobility, despite its size.

Firepower

Produced by KUKA of Augsburg, the two-man forward-mounted turret

MECHANISED COMBAT

Inside the Marder

Entering service in 1971, the Marder was a far more capable vehicle than the M113 APCs widely used by the Bundeswehr. Compared with the Soviet BMP series, the Marder was large and lacked anti-tank capability, but has now been fitted with MILAN.

Commander
The commander and gunner both have PERI Z11 sights offering ×2 and ×6 magnification. The Marder A1 is fitted with an image intensifier.

Co-axial MG3 7.62 mm machine gun
Marder carries 5,000 rounds of 7.62-mm ammunition for the co-axial and remote controlled machine-guns.

Smoke dischargers

Gunner
The 20-mm cannon is served by three separate belts so the gunner can rapidly alter his choice of ammunition as different targets appear.

Driver
The driver has three periscopes, one of which can be replaced by a night vision device.

Engine compartment
The MTU MB 833 six-cylinder liquid-cooled diesel developes 600 hp at 2300 rpm.

is among the most advanced of its type. The commander and gunner, mounted on the left and right respectively, each has a single-piece hatch-cover and adjustable seat. Turret traverse and gun elevation are operated electro-hydraulically, whilst loading and unloading, cocking, firing and reloading are all executed under armour protection.

The 20-mm Rheinmetall Mk 20

Marder at speed with the commander clinging to the rim of his hatch. Despite being two or three times as heavy as many APCs, the Marder has a good power-to-weight ratio and excellent battlefield mobility.

95

MECHANISED COMBAT

Having backed his vehicle into the trees, this Marder commander prepares to fire MILAN. All German army Marders are now fitted with MILAN except their command vehicles.

Rh-202 cannon is fed via a series of rigid and flexible chutes from three different belts to give the gunner a choice of either armour-piercing (AP) or high-explosive (HE) shells.

A dual control system enables the commander to over-ride the gunner in the case of an emergency, whilst the turret itself can be operated manually via a series of mechanical gear-boxes and the gun fired by foot controls in the case of a failure in the hydraulic or electrical systems. To avoid fumes and clutter, empty cartridge cases are ejected automatically outside the turret.

Since 1982 most Marders have been improved by the retro-fitting of a double belt feed for the 20-mm cannon, improved night capabilities and an enhanced image intensifier with thermal pointer.

A MOWAG-designed remote-controlled 7.62-mm MG3 machine-gun mounted above the rear of the crew compartment gives the Marder a unique sting in the tail.

With the exception of command vehicles, all Marders in German service are fitted with a Euromissile MILAN ATGW launcher.

Variants

A number of early variants of the Marder were cancelled either as being ineffective or too expensive and were subsequently replaced by derivations of the less complex and far cheaper United States' M113 APC. Others were, however, proceeded with and are presently operational.

After strenuous competition between Rheinmetall and Mauser for the

A Marder with hatches open: the position behind the driver is occupied by one of the infantrymen, who dismounts to fight. The height of the Marder (nearly 3m [10ft]) is readily apparent. The Russian BMP was just 2.15m (7ft).

production of a new 25-mm cannon to replace the present 20-mm Rheinmetall system, the latter proved successful and has recently been awarded a development contract to retro-fit their Mauser E into the existing KUKA turret. In addition to firing HE and AP ammunition, the Mauser will accept an armour-piercing fin-stabilised discarding sabot (APFSDS) round capable of penetrating the armour of the latest Russian BMP. Although, due to financial constraints, the new cannon will not become fully operational for several years, its very existence guarantees the continuation of the Marder into the 21st century.

A small number of Radarpanzer TURs (Tiefflieger Uberwachungs Radar) entered service in 1981. Consisting basically of a Siemens radar with a range of some 30 km (18.5 miles) attached to a hydraulically operated arm, the base of which is itself welded to an extensively modified turretless Marder chassis, the TUR has a crew of four and is armed with two 7.62-mm machine-guns for local defence.

The Bundeswehr also uses the Marder chassis to mount the highly successful Euromissile Roland 2 surface-to-air missile developed jointly with France. Two missiles are carried in the ready-to-use position, with a further eight stowed internally. With a maximum range of 6,300 m (20,670 ft), a minimum range of 500 m (1,640 ft) and radar range of 18 km (11.2 ft), the Schulzenpanzer Roland, of which 140 examples are in service, provides an excellent companion to the twin-barrelled 30-mm Gepard, itself one of the best anti-aircraft gunnery systems in NATO service.

Argentine tank

To meet Argentinian requirements, Thyssen Henschel has developed the TAM (Tanque Argentino Mediano) medium tank, of which 300 have recently been built under licence. Equipped with a more powerful 720-hp engine, less sophisticated 20-mm cannon and a 7.62-mm anti-aircraft machine-gun, the TAM is in most other respects similar to the Marder and is itself proving a great success as a 'parent' vehicle for a large series of variants.

Since 1982 most Marders have been upgraded to either A1 or A1A status. Stowage and storage have been improved, flaps fitted to the periscopes and the commander's external NBC system for utilisation when firing the MILANs. The crew of four has been retained but the number of passengers has been reduced to five.

Despite its age, Marder is still an excellent fighting vehicle and remains the equal of such later systems as the British Warrior and United States Bradley. It will clearly play an important role in German military strategy for many years to come.

The infantry fires G3 rifles from their roof hatches. They can also fire from within the vehicle using the firing ports in the hull side. A ventilation system clears away the fumes far more efficiently than in the Russian BMP.

MECHANISED COMBAT

MOBILE OPERATIONS

OPERATING FROM VEHICLES

For safe operations, remember the following:
1. When deployed in close country, you must have some infantry deployed on foot to protect you.
2. Use scout groups and flank protection where possible.
3. Every vehicle must carry spares and the crews must be trained to carry out repairs.
4. Vehicles must be standardised as far as possible so that damaged vehicles can be cannibalised.
5. Plan for POL (petrol, oil and lubricants) consumption of two to four times the normal rate.
6. Commanders must not travel in the same vehicle.
7. Everyone must know the anti-ambush and anti-aircraft action.
8. Every vehicle commander must map read rather than play 'follow the leader' and they should all know the emergency RV.

During ATOPS there is the ever-present danger of vehicles being ambushed by terrorists. The risk varies depending on the nature of the terrain and enemy activity.

Generally, the cities and towns are fairly secure. The danger increases the further one travels from built-up areas. At the height of the Rhodesian War, civilian traffic was seriously disrupted by the ever-present threat of terrorist ambush. If you wanted to travel from one town to another, you had to join a protected convoy that would form at a certain time at a designated spot. Such convoys stuck to main roads. It was wise to steer clear of the lonely, unmetalled roads, which were often mined.

In an attempt to discourage terrorist activity, the Rhodesians placed picquets along all roads leading in to and out of major towns. As the war progressed, so did local ingenuity. Special vehicles were designed. These armour-plated, often heavily armed monstrosities were affectionately named after local animals, such as the 'Croc' and 'Hippo'.

The tracker unit having sighted the enemy, the rest of the unit deploy from their Casspirs to follow up on foot. When moving on roads, vehicle crews would normally ride on top of the vehicle: it left you more vulnerable to small-arms fire, but less likely to be severely injured by mine attack. Note the very mixed uniform of the South West African Police Counter Insurgency unit.

A South African Mirage on a ground attack practice mission fires its rocket pods. In the past, large SWAPO convoys were effectively targeted for air strikes by South African special forces manning observation posts in Angola.

MECHANISED COMBAT

Attacks on civilian vehicles became more and more frequent. The Army provided protection for convoys travelling between towns. The truck is the Rhodesian version of the Unimog. Note the RPK mount.

Preparation prior to moving off for Operation Uric in Mozambique. The 90-mm rounds for the Eland's main armament can be seen through the open hull door. The support vehicle carries a generous supply of spare tyres.

The floors of heavy vehicles were covered in a compact layer of sandbags as some protection against mines. In convoy, vehicles had to maintain both radio and visual contact, and firepower was evenly distributed throughout. Troops travelled in open-topped trucks, sitting back to back, facing outwards, with weapons cocked.

If a convoy was ambushed, the greatest danger came from halting within the killing ground. You should stop *before* running into the ambush area or, if this is impractical, *you must keep going until clear*. As soon as you are attacked you must return fire and de-bus the moment your vehicle stops. You should then mount an immediate counter-attack from flank to rear.

If you *are* forced to halt within the killing ground, you should instantly de-bus and launch a counter-attack. To be ambushed is a terrifying experience. Don't be tempted to take cover in the monsoon ditch along the edge of the road, however! The ditch, and any obvious cover will probably be mined. Your only chance of survival is immediate, aggressive, reaction.

Above: A convoy halts during its return from Operation Uric in Mozambique. Unlike the South African equivalents, these vehicles are very vulnerable to small-arms fire and mines. ZANLA and ZIPRA became skilled in the use of simple but effective booby traps which made all movement by road hazardous.

Left: South Africans equipped with Casspirs motor through the bush on a follow-up operation. The vehicles provide good protection from mine blast and small-arms fire.

MECHANISED COMBAT

Left: Cross-border raid convoys can always call in an air strike from aircraft such as this Buccaneer, if they find themselves in trouble or get ambushed.

Right: A convoy of former Rhodesians on the route back from a visit to Mozambique. The trucks are 2.4-tonne (2.5-ton) Unimogs. The mortar truck in the foreground is capable of carrying full first-line ammo, 106 HE bombs and 48 smoke bombs. All the vehicles are also fitted with MAG GPMGs.

Below: The results of a mine on a Rhodesian truck. Fortunately, the driver and crew survived without serious injury. However, enemy mining succeeded in effectively isolating large areas of the countryside from the security forces in Rhodesia.

Ambushing terrorists

The aim of an ambush is to surprise and eliminate the enemy on ground and in circumstances of your choosing. Ambushes can vary in size from a small, four-man affair, to a major operation involving a platoon or company group. The key to a successful ambush is instantaneous co-ordinated action, against a surprised enemy, within a well-covered killing ground.

Three fundamental principles govern the ambush layout:
1 All possible approaches should be covered
2 A killing ground must be carefully selected
3 The ambush must have depth

Types of ambush depend on a number of factors. Intelligence may dictate a **limited ambush** at a specific point. If the exact spot is not known, an **area ambush** should be laid to cover all possible enemy approaches and routes. An ambush can be either a short-term or a long-term operation, with the latter lasting anything from nine hours to several days. Obviously, the longer you are in position, the more your chances increase of being compromised.

Immediate ambush

Occasionally, an opportunity will present itself for an **immediate ambush**. An excellent example of this occurred in Angola in August 1981, when the Pathfinder Company of 44 Parachute Brigade (SADF) successfully hit an enemy convoy that had unwittingly halted immediately in front of their night-time position:

"The vehicles... nearest the enemy opened up with .50 cals, MAGs and the 60-mm mortar. The FAPLA convoy replied with small arms and a number of 23-mm twin-barrelled anti-aircraft machine-guns... After the firefight had raged for about 15 minutes, the enemy's fire ceased..."

A hastily formed patrol found the enemy setting up mortars. They were quickly dealt with, and FAPLA hastily withdrew, leaving behind 14 vehicles.

Mines and booby traps have become a hallmark of terrorist activity worldwide. Types of device used by

MECHANISED COMBAT

Above: Casspir armoured personnel carriers were used to patrol vast stretches of bush in Namibia. The guerrillas used Polish anti-tank grenades fired from AK-47s against them.

Above: Soldiers of the South West African Territorial Force were recruited by the SADF. Many had to flee to South Africa after SWAPO came to power.

the enemy can be divided into two categories – non-explosive and explosive.

Non-explosive booby traps, such as vehicle pits, lassoes, nets and so on, are unusual but you may still encounter them. In themselves, however, such traps are not generally lethal and may therefore be accompanied by an ambush.

You are more likely to come across mines and explosive booby traps. The most common types of mine are the anti-vehicle or anti-tank, anti-personnel, or improvised mine. Most mines consist of five essential components: the initiating action; trigger mechanism; detonator; primer; and principal charge.

Mines can be initiated by pressure (for example, the Russian TM-46 AT mine); completion of an electrical circuit (as in the Claymore); when a pull is exerted (as in the Russian POMZ-2/2M AP mines), or by a combination of methods.

A mine can be concealed in almost any kind of terrain, and not necessarily underground! It can just as easily be hidden in a bush or tree. A wily terrorist can also lead you onto a mine simply by placing an obstruction across your path, thereby forcing you to make a detour.

Mines may be laid singly, or have their power increased by having one or more 'booster' mines placed on top of the other. They can also be interconnected. Almost all come with, or are easily fitted with, an anti-lifting device.

Explosive booby traps can be initiated by a variety of means, depending on the material to hand, the type of trap required and time available to set them up. Traps vary from the complex and sophisticated to the simple, yet equally functional. A grenade with a shortened fuse, pin out and lever held down by a dead weight – although something of a cliché – when employed in the right circumstances will almost always account for at least one unsuspecting victim.

Protecting sensitive points

Sentry duties are hardly the most exciting of tasks, but are nevertheless a vital aspect of bush warfare. You must remain alert and vigilant and carry out this task as you would any other – or pay the consequences, as ZANLA and ZIPRA often did ... but then, they were up against the most professional fighting force ever to emerge in southern Africa!

Crew members clean the 90-mm gun of an Eland armoured car belonging to the South African Defence Force. The other version of the Eland has a 60-mm mortar instead. The Eland is a locally-modified version of the French AML-90 design.

MECHANISED COMBAT

PANZER RECCE!
THE SPÄHPANZER LUCHS

The modern commander has an infinite variety of intelligence-gathering agencies available to him. Many of these, notably the spyplane, satellite and electronic countermeasure systems, are modern and highly complex. Others, personified by the infantry recce patrol, are as old as the army itself.

The German army introduced a series of revolutionary scout cars before World War II. Relying on speed rather than armoured protection, and with just enough firepower to extricate themselves from danger, these vehicles acted as the eyes of the Panzer Divisions. As the war progressed and the Allies began to construct their own, they followed a similar pattern with a four-wheel-drive chassis, armoured hull and rotating turret mounting a light-calibre quick-firing weapon. The modern reconnaissance vehicle typified by the Luchs is a direct descendant of these armoured cars.

Development of the Luchs began in 1964 when the German Ministry of Defence began to implement plans for the creation of a new family of military vehicles to include an 8×8 amphibious reconnaissance vehicle, 4×4 and 6×6 amphibious load carriers and a series of trucks. Daimler-Benz and a consortium led by Krupp, MAN and Rheinstahl began the immediate development of rival prototypes. In 1973 a contract for the production of the Luchs reconnaissance vehicle was awarded to Rheinstahl. Production began in May 1975 and by the end of 1977 408 models had entered service.

The all-welded hull of the Luchs is proof against small arms fire and shell splinters, although its comparatively thin armour would afford little protection to the four-man crew against a direct hit from the weapon systems of the latest generation MICVs.

Above: The large size of the Luchs is readily apparent when all four crew members have their hatches open. Note the radio operator/rear driver at the back: he has a complete set of controls and can drive the vehicle swiftly out of trouble.

Room with a view

The driver seated at the front-left of the hull is provided with a single-piece hatch cover opening to the right and is equipped with three periscopes offering reasonable forward vision even when driving closed-down. Uniquely a second driver, who also acts as the radio operator, is situated

A column of Luchs armoured cars assembles for action. Note the door in the hull side between the two sets of wheels and the propellers on the hull rear, which power the Luchs at up to 9km/h (5.6 mph).

MECHANISED COMBAT

The Luchs suspension system

vertical spring

axle

To allow the Luchs to bounce across rough ground at a suitably vigorous pace, the suspension system consists of four rigid axles supported by longitudinal bars. Each wheel station has a vertical coil spring and hydraulic shock absorber.

Inside the Luchs

The Luchs is directly descended from the excellent 8×8 armoured cars used by the German army in World War II. The requirement to be fully amphibious led to a large hull, which makes the Luchs one of the largest recce vehicles in service. Production has been completed and the Luchs has not been exported.

7.62-mm machine-gun
A Rheinmetall MG3 GPMG is fixed to a skate mounting above the commander's hatch. Its value as an anti-aircraft weapon is rather limited, and to engage enemy infantry the commander would have to make himself vulnerable to small-arms fire.

Gunner

Driver's periscopes
The central periscope can be replaced by a passive night periscope.

Driver

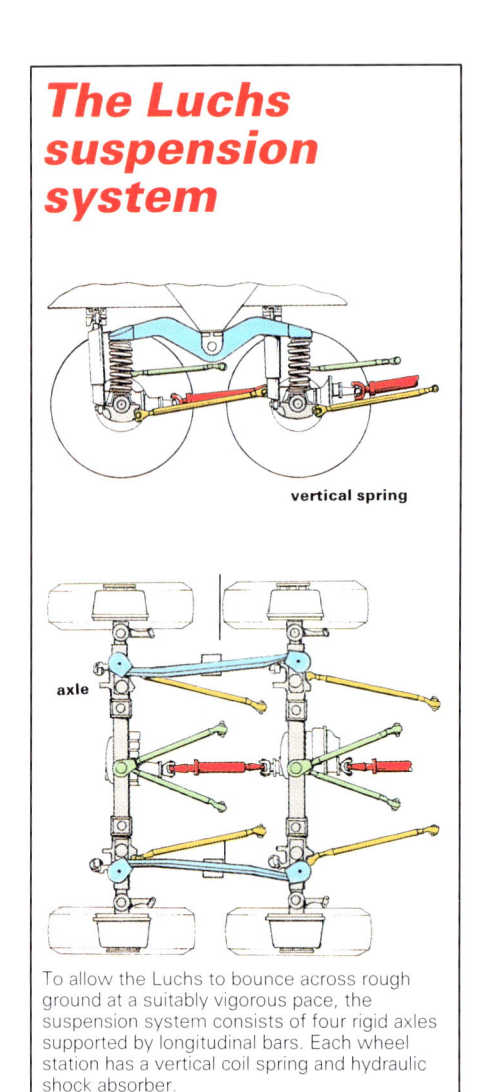

Hull frontal armour
This is proof against weapons up to 20-mm calibre: the primary Russian opponent is the BRDM armoured car which carries a 14.5-mm machine-gun.

Trim vane
This is erected hydraulically before the Luchs enters the water.

facing rearwards behind the engine. Equipped with a similar configuration of single-piece hatch cover and three periscopes, when directed the rear driver is capable of taking over complete control of the vehicle.

Whether the obvious theoretical advantages of a second driver are as apparent in reality is a matter for debate. In action the rear driver will be constantly attending to the stream of signals associated with a reconnaissance vehicle once its presence has been compromised and he is more than likely to be severely disorientated; a problem well known to mechanised infantrymen operating from the rear of enclosed armoured personnel carriers! It is therefore highly unlikely that he will be able to assume immediate responsibility for

A luchs heads across still water with trim vane erected. Like the Russians, the Germans believe recce vehicles should be truly amphibious and thus able to cross a water obstacle without elaborate preparations.

MECHANISED COMBAT

Smoke rocket dischargers
Four are mounted on each side of the turret.

Turret armour
This is proof against weapons up to 20-mm calibre all round.

Searchlight
This can be used in infra-red mode and it follows the same elevation as the 20-mm cannon.

20-mm Rheinmetall Mk 20 Rh202 cannon
This is the same weapon as mounted on the Marder MICV. It uses a dual feed and ejects the empty cases and cartridge links to the right of the turret. It provides a hail of anti-personnel fire, but its value against armoured vehicles is limited.

Radio operator/rear driver
The unique feature is the second driving station in the hull rear, which allows the radio operator to get the vehicle out of trouble fast. The Luchs can travel backwards as fast as it goes forwards, which is very handy in the cut-and-thrust world of armoured recce.

Hull side armour
The side and rear of the Luchs are sufficiently armoured to keep out small arms fire and shell splinters, but can be penetrated by heavy machine-gun or cannon fire.

Hull door

Below: Russian BMP armoured personnel carriers. With 15–20 mm (0.6–0.8 in) of armour, the BMP could have been a dangerous opponent for the Luchs, which would have had to rely on its speed to break contact.

charting a rearward passage to safety without considerable assistance from the commander. Well trained crews of conventionally driven reconnaissance vehicles practise the art of speedy withdrawal incessantly, with the result that an experienced commander-driver team can reverse their vehicle immediately and at speed even when the latter is totally unsighted.

The engine, transmission, air filter and oil filter, located together at the rear of the turret, are mounted as a complete unit to facilitate speedy removal from the vehicle under field or other adverse conditions. The engine

MECHANISED COMBAT

A Luchs armoured car probes cautiously towards the edge of a wood during a NATO exercise. An 8x8 all-wheel driven vehicle, the Luchs is an exceptionally mobile AFV.

employed is the well proven Daimler-Benz Type OM 403 VA, 10-cylinder, four-stroke with fuel injection and turbo-charger. Capable of developing 390 hp with diesel fuel and 300 hp when used with petrol, this thoroughly reliable engine enables the Luchs to attain a top road speed of 90 km/h (56.25 mph), both forwards and backwards, and to accelerate from 0/80 km/h (0/50 mph) in 65 seconds, a considerable feat for a vehicle that weighs 19,500 kg (42,990 lb).

The Luchs is fully amphibious with the aid of two propellers, one on each side of the hull rear, being capable of a top speed in current-free water of 10 km/h (6.25 mph). Preparation is negligible, a trim vane being erected hydraulically by the driver before the vehicle enters the water. Steering, effected by swivelling the propellers, is adequate but not excellent.

Aided by the driver's ability to select either the front four wheels, rear four wheels or all eight wheels for steering, the Luchs has an excellent cross-country capability. With a maxi-

A Luchs demonstrates its splendid suspension system. Although all eight wheels are steered, only the front four are normally used when travelling along a road. The vehicle's armour is necessarily thin but at least the front of the hull is very well sloped.

mum road range of 800 km (497 miles), coupled with the capacity to climb gradients of 60 per cent, traverse trenches 1.9 m (6.2 ft) deep and overcome vertical obstacles of 0.6 m (2 ft), it remains, despite its age, one of the most versatile reconnaissance vehicles in current service.

Staying out of trouble

The Luchs is in every respect a reconnaissance vehicle and as such has not been designed to enter into fire fights. The two-man power-operated turret is situated behind the driver. The main armament is a 20-mm Rheinmetall Mk 20 Rh 202 cannon, identical to that mounted on the Marder MICV. Capable of accepting high-explosive or armour-piercing shells, the gun is extremely accurate and reliable. However, the Luchs lacks the sophisticated automatic loading system found in the Marder and as such, unless it were to attain a first round kill, would be no match for the latest generation of Russian armoured vehicles, such as the BMP-2. A Rheinmetall 7.62-mm MG3 machine-gun is mounted on a ring mount over the commander's hatch.

The Luchs is fitted with an NBC system and automatic fire-extinguishing system, as well as a highly versatile heater capable of blowing hot air over the engine if required. Unlike many German vehicles of its generation Luchs was never exported, nor were any variants produced. Nevertheless, 408 models remain in Bundeswehr service and show no signs of being replaced.

MECHANISED COMBAT

ELAND IN ACTION

You will encounter the French Panhard AML series of armoured cars all over the world. Used by over 35 different armies, the AML has been manufactured under licence in South Africa and has seen a great deal of action in Namibia and Angola. The vehicle was designed after the French army's experience with British Ferret armoured cars during the Algerian war, the first AMLs entering service with the French army in 1961 shortly after the abortive coup led by the Foreign Legion paras.

The AML 90 is fitted with a Hispano-Suiza turret. The commander

The Eland was developed in South Africa from the French AML light armoured car. It is fitted with a 90-mm gun, which will fire HEAT rounds accurately to at least 1,000 m (3,280 ft). Although not very well protected, Elands proved to be a match for heavier enemy tanks during the long war on the Namibia/Angola border.

and the gunner both have four periscopes; the latter also has an M37×6 magnification sighting periscope. Doors are situated on both sides of the hull beneath the turret ring. The spare wheel is concealed in the compartment behind the rearward opening left door, and the small area behind

MECHANISED COMBAT

The alternative weapon for the Eland is the 60-mm mortar. There is a 20-mm cannon turret which has recently been trialled and is now being retro-fitted to the older models.

the right door provides the crew with its only storage area for personal kit and equipment as well as brew-kit and rations – a perennial problem in all armoured reconnaissance vehicles.

The main armament is a 90-mm D921 Fi gun which can fire HEAT (high-explosive anti-tank), HE, smoke or canister rounds and is aimed with the assistance of a 7.62-mm co-axial machine-gun positioned to its left. A further 7.62-mm (or occasionally a 12.7-mm) anti-aircraft machine-gun may be mounted on the roof. It is not stabilised and therefore ineffective on the move, but the gun is capable of a rapid rate of fire, can traverse through 360° in 20 seconds and can be elevated from +15° to –8°. Twin SS-11 or ENTAC ATGWs were offered with

This is the blast of the 90-mm night-firing. The Eland can certainly hand out punishment but is not good at taking it, as 15 mm of front armour will only stop 7.62-mm small-arms fire.

early models, but were not effective and were withdrawn.

The Lynx-90 turret

Hispano-Suiza has developed an excellent Lynx-90 turret suitable for retro-fitting to a number of French-produced armoured cars, including the Panhard AML. Of all-welded steel construction, the turret, with its front armour of 15 mm (0.6 ft) is proof against small-arms fire, but offers little protection against the range of ammunition mounted on the majority of modern MICVs (Mechanised Infantry Combat Vehicles). The commander (seated on the left) and the gunner (to his right) both have single-piece hatch covers and adjustable seats. In common with French Main Battle Tanks, the commander has a raised rotating cupola for greater all-round observation, which can pintle-mount a 7.62-mm machine-gun for anti-aircraft defence.

In 1983 the much improved Lynx 75/90 turret, capable of mounting the 90-mm Cockeriil Mk III gun, the Thompson Brandt breech-loaded mortar and a variety of smaller 75-mm guns, was introduced. Power controlled with manual override, it can turn through 360° in only eight seconds and so can engage enemy forward reconnaissance vehicles before they have the chance to retaliate. Vehicle commanders know that at short ranges such vehicles rarely miss, so the ability to strike the enemy first is a great comfort.

The Eland

In the early 1960s South Africa obtained a licence from Panhard to undertake production of the AML. Sandock-Austal Beperk Limited of Boksburg in the Transvaal, produced several thousand examples, of which 1,600 are currently in service. Designated the Eland, the original design

Inside the AML 90

The AML 90, as far as export sales are concerned, is one of the most successful armoured vehicles in history. It is cheap, robust, easy to produce and easy to update with new technology. This combat-proven vehicle is likely to be in service well into the next century.

Main armament
The 90-mm D921 F1 gun is fitted in an H 90 turret originally manufactured by Hispano-Suiza in the earlier models. Later models are fitted with the Lynx 90 turret shown here. The gun can fire HEAT, HE, smoke and canister rounds for use against infantry in the open. All the rounds are fin-stabilised and the HEAT round will penetrate 320 mm (12.6 in) of armour at 90 degrees to the plate and 120 mm (4.7 in) for a glance angle of 65 degrees.

Driver's single-piece hatch cover
This opens to the right and has three integral periscopes. The centre scope can be replaced with an infra-red or image intensification periscope for night driving.

Driver

Sand mats

All-welded steel hull
This is divided into three compartments: the driver's compartment, a fighting compartment in the centre, and the engine compartment at the rear.

incorporated a 60-mm mortar platform and twin 7.62-mm machine-guns. Since then, numerous improvements have been incorporated as a result of combat experience in Zimbabwe (a few models were exported to the Smith regime prior to independence) and Namibia. The latest models have a powerful petrol engine specially geared to the hot, dry climates of the bush and a new transmission, which can be changed even in field conditions in 40 minutes.

The final production models were armed with a 90-mm gun, 7.62-mm co-axial and 7.62-mm anti-aircraft machine-guns. Well-trained South African crews proved that the Eland was more than a match for enemy T-34 and T-54/55 tanks in Angola.

Easy to produce and 95 per cent

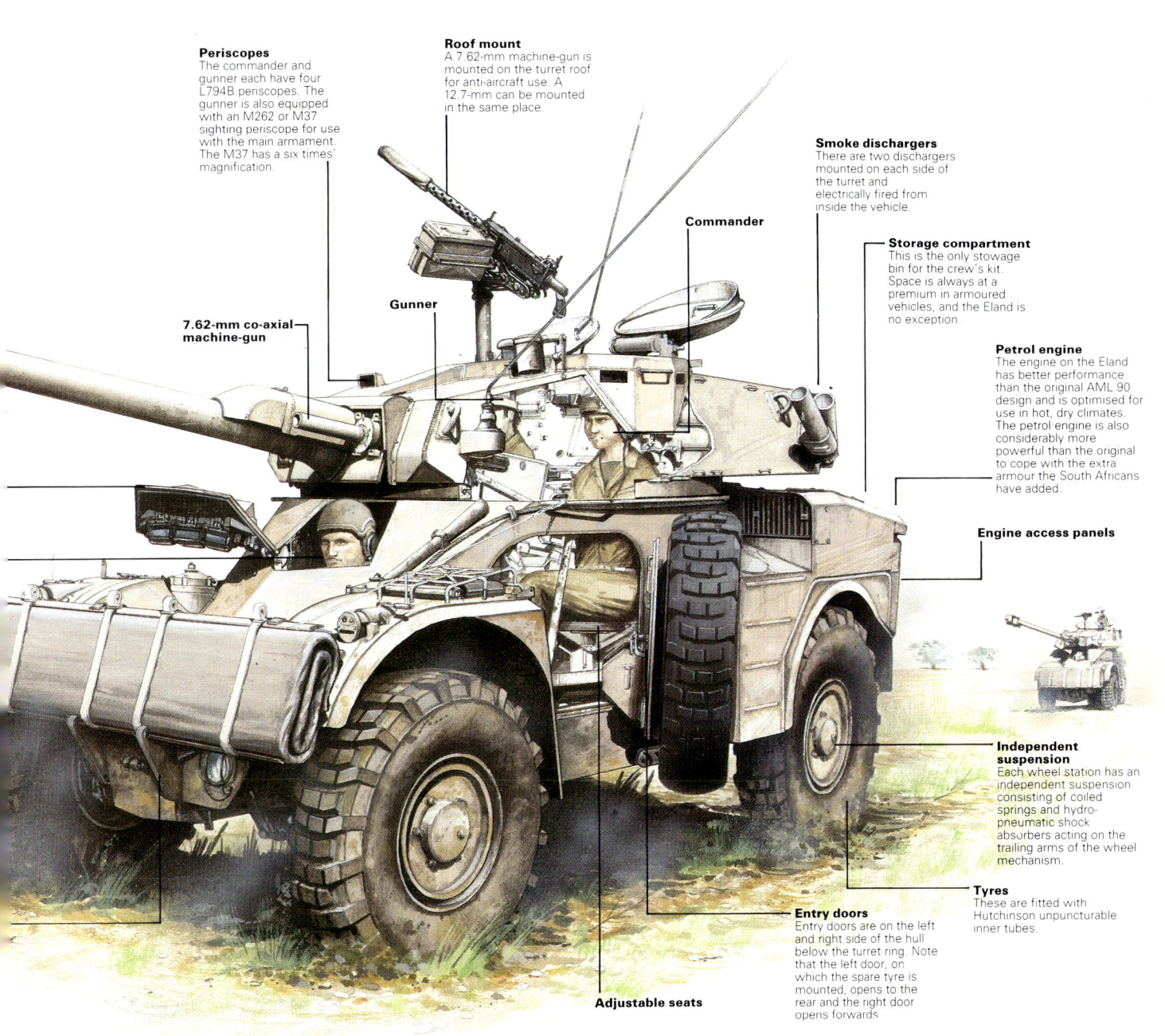

MECHANISED COMBAT

Periscopes
The commander and gunner each have four L794B periscopes. The gunner is also equipped with an M262 or M37 sighting periscope for use with the main armament. The M37 has a six times' magnification.

Roof mount
A 7.62-mm machine-gun is mounted on the turret roof for anti-aircraft use. A 12.7-mm can be mounted in the same place.

Smoke dischargers
There are two dischargers mounted on each side of the turret and electrically fired from inside the vehicle.

Storage compartment
This is the only stowage bin for the crew's kit. Space is always at a premium in armoured vehicles, and the Eland is no exception.

Petrol engine
The engine on the Eland has better performance than the original AML 90 design and is optimised for use in hot, dry climates. The petrol engine is also considerably more powerful than the original to cope with the extra armour the South Africans have added.

Engine access panels

7.62-mm co-axial machine-gun

Gunner

Commander

Independent suspension
Each wheel station has an independent suspension consisting of coiled springs and hydro-pneumatic shock absorbers acting on the trailing arms of the wheel mechanism.

Tyres
These are fitted with Hutchinson unpuncturable inner tubes.

Entry doors
Entry doors are on the left and right side of the hull below the turret ring. Note that the left door, on which the spare tyre is mounted, opens to the rear and the right door opens forwards.

Adjustable seats

domestic, the Eland is an excellent weapon system for South Africa. A two-man turret capable of mounting a 20-mm cannon and co-axial 7.62-mm machine-gun has been developed in recent years and has been retro-fitted to a few existing chassis. The Eland has given magnificent service to South Africa for over four decades, and will continue to form a crucial part of the SADF's equipment for many years to come.

Eland crews who are providing flank protection for the convoy on the road clean the main armament prior to moving out. The Eland has good cross-country performance and the new petrol engine gives an excellent power-to-weight ratio.

107

MECHANISED COMBAT

In the best traditions of French design, the AML is capable of accepting several different turrets and weapon systems, of which the most popular is based on the Hispano-Suiza HE-60 series. Two initial models, the H60-7 and H60-12, are each equipped with a Thompson Brandt 60-mm mortar supported by twin 7.62-mm machine-guns and a 12.7-mm heavy machine-gun respectively. Despite its comparatively small size, the two-man all-welded cast turret can hold 43 rounds of mortar ammunition and either 3,800 rounds of 7.62-mm or 1,300 rounds of 12.7-mm. It is fully traversable in 25 seconds and capable of a motor traverse of −15° to +80°, making it a highly effective Third World weapon system. A later variant, the Serval 60/20, entered service in the mid-1980s and proved an export success. It can accept the HB 60 60-mm mortar capable of a direct fire range of 1,000 m (3,280 ft), supported by a 20-mm cannon mounted externally at the rear of the turret.

A number of smaller, less complex scout car variants ideal for the African and Far Eastern climates exist, including the EPR scout car with a single 12.7-mm machine-gun, the EPF border protection vehicle and EPA airfield protection vehicle. The last is designed to carry a cache of 50 hand-grenades as well as a fearsome array of three 7.62-mm machine-guns.

As an export vehicle, the Panhard AML has been one of the most successful creations in armoured history. Cheap and versatile, it has served the French army well and the friends of France better. There can be no doubt that the AML will be in service well into the 21st century.

This South African Defence Force armoured patrol is pictured 100 km (62.5 miles) inside Angola after a successful 1983 raid on SWAPO guerrilla bases in the country.

Later versions of the Eland were a considerable improvement on the original AML, having more armour, an engine optimised for hot weather operations, and a transmission that can be changed in under 40 minutes.

MECHANISED COMBAT

BREAK ON THROUGH WITH THE 432

The FV 432 armoured personnel carrier is used exclusively by the British Army. Three thousand were built between 1962 and 1971. Warrior will replace them as APCs but FV 432s will have to soldier on for many years in other roles.

Few weapon systems in history have served the British Armed Forces as well as the FV 432 series of mechanised armoured vehicles. Designed originally in the early 1950s, the design was shelved for several years until the need to update British infantry battalions based in West Germany from motorised to mechanised became more pressing with the advent of new generations of Soviet armoured personnel carriers. Research continued, resulting in the completion of the first FV 430 prototype in 1961.

In 1962 GKN Sankey was awarded a production contract for the FV 432. The first production vehicles were completed in 1963 and by the time production was completed in 1971 over 3,000 models had been manufactured; 2,300 remain in active service today.

Design

The all-welded steel hull of the FV 432 is simply an armour-plated steel box on tracks, with an engine at the front and crew compartment at the rear. Complete protection is afforded the crew and passengers against small arms and shell splinters, although it is impossible for the commander to fire the 7.62-mm GPMG mounted on the forward part of his cupola without first exposing his head and shoulders.

The driver, who is seated at the front to the right of the engine, has a single-piece hatch opening to the left. When driving closed-down with his hatch cover secured, his visibility is severely curtailed by the provision of a single wide-angle periscope.

The commander, seated directly behind the driver, enjoys better visibility. His cupola, which can be traversed manually through 360 degrees, has a single-piece hatch cover and three periscopes.

It is usually stated that the crew compartment can hold an enlarged

109

MECHANISED COMBAT

Assaulting an enemy-held position on exercise: a Chieftain MBT provides heavyweight close support ahead of the infantry who have dismounted from their FV 432s. The nearest 432 has a Peak Engineering turret mounting a GPMG.

Frontal armour
The glacis is protected by 12mm of armour. The hull is all-welded steel, which largely accounts for the FV 432 being substantially heavier than the aluminium M113. This makes the British vehicle a little tougher but no more able to withstand a hit from an anti-tank weapon.

Wide angle periscope
Fitted to the driver's hatch, this periscope can be replaced with a passive night vision periscope for night driving.

Air intake

Drive sprocket

Torsion bar suspension
The FV 432 has five dual rubber-tyred wheels, the first and last of which are fitted with a friction shock absorber.

Top track cover
This light sheet metal covering is often removed to make cleaning and maintenance easier.

Rubber bushed steel tracks

Inside the FV 432

The FV 432 belongs to the generation of APCs designed during the 1950s which followed the 'armoured box on tracks' approach. Similar in concept to the American M113, the British vehicle is bigger, heavier and slower, and not amphibious.

section of 10 fully equipped infantrymen seated on bench seats running down each side of the hull. However, in practice a normal section of eight men is cramped and uncomfortable if confined for too long. In simulated Nuclear, Biological and Chemical (NBC) attacks whole sections are occasionally ordered to remain in their vehicles for hours, if not for days, on end. Their sheer relief at 'endex' is easy to imagine!

Normal means of entry and exit for the infantry is via a single large door at the rear. Ideally this opens easily to enable speedy entry and exiting, but if the vehicle stops at an angle, opening and closing of the heavy door can be difficult and time-consuming. In an emergency the troops can also use the four-piece hatch above the crew compartment, but this was designed as a mortar fire point and would leave the exiting troops very vulnerable to enemy fire.

Powerpack

The Rolls-Royce multi-fuel engine is coupled to the General Motors semi-automatic transmission with a choice of six pre-selected gears to give the FV 432 a maximum road speed of 52 km/h (32.3 miles), at a range of 380 km (236 miles) and sufficient power to climb gradients of 60 per cent and vertical obstacles of 60 cm (24 in).

To facilitate repairs in the field, the complete engine with its oil tanks and filters is mounted on a common subframe which can be lifted out of the vehicle in one piece with the aid of the crane mounted on the FV 434 engineering vehicle. If conditions permit, the engine can then be reconnected on the ground with the aid of extended cables and fuel lines and repaired in comparative safety and comfort.

Suspension

The torsion bar suspension gives a tolerably even ride for the passengers in their cramped quarters. However, the high-pitched whine of the engine coupled with the complete enclosure of the troop compartment when closed down can lead to serious disorientation among the infantry immediately after disembarkation.

An NBC system mounted on the right-hand side of the hull provides fresh air ducts to the troop and driver's compartments.

Unlike the United States' M113 the FV 432 is not amphibious without preparation. To swim, a flotation screen must be fitted, held on by 10 stays; a trim vane is erected at the front of the hull and an extension fitted to the exhaust pipe. When afloat the vehicle is unwieldy and highly susceptible to river currents: as a result, most vehicles have now had their

FV 432s have a well designed GPMG mount in front of the commander's cupola. When the vehicle is stationary this is almost as accurate as the GPMG in the Sustained Fire role.

MECHANISED COMBAT

Driver's hatch
The driver sits on the right-hand side of the hull with the engine to his left.

Commander's hatch

Commander
He has a manually traversed cupola with 360 degree traverse and three periscopes. The excellent GPMG mount allows him to shoot with great accuracy when the vehicle is stationary, but he is completely exposed while doing so.

Engine decking
The rear part of the decking is the air outlet and gets extremely hot when the FV 432 is on the move. If you are standing in the troop compartment with the top hatch open, watch where you put your hands.

Top hatch
The circular hatch above the troop compartment opens into two hatches which can be locked open. For clarity, the left-hand hatch is not illustrated.

Rear door
The decision to have a door rather than a ramp, as fitted to the M113, has mixed blessings. It is fine on level ground, but operating it while the vehicle is on a slope requires time and strength.

Boiler
Forget the armour and the mobility: what makes the FV 432 good news for the infantryman is the small boiler in the door. No need to fiddle with hexamine blocks; you can get a brew on in no time at all. Tinned food can also be cooked here, and if there is not enough room for all the cans you can pierce them and wedge them between the hull and exhaust pipe.

Idler

Exhaust pipe

amphibious capability removed.

The 24V negative earth electrical system is current-rectified for battery charging and general purposes. A small but very useful water heater is provided in the passenger compartment, which when used properly ensures that all aboard can enjoy not only the luxury of almost constant hot drinks but also regular hot meals – an essential in combat conditions.

*The troop compartment of an **FV 432** with an acquired portable bench on the left. Prolonged **NBC** exercises in which a section sits for several days in the troop compartment have revealed the virtual impossibility of keeping men fit for action in this way.*

111

MECHANISED COMBAT

The ambulance version carries four stretcher patients or two stretcher and five seated wounded. Sliding swivel sling racks in the rear allow the stretchers to be loaded and unloaded quickly.

Entering a freshly-painted FV 432 during a public display: officially the troop compartment can accommodate 10 soldiers. In reality, eight men and their equipment fill the vehicle to capacity.

Variants

The FV 432 was designed specifically as one of a series of new mechanised vehicles, so numerous variants exist.

The most striking variant, the FV 434 Armoured Repair Vehicle, is operated by the Royal Electrical and Mechanical Engineers (REME) usually far forward in the battle zone, where it is used to repair disabled and damaged vehicles. Strangely, the FV 434 has no recovery capability, members of the FV 432 family being recovered where necessary by standard APCs fitted with a double capstan winch attached to the rear of the chassis. The HIAB crane fitted to the right of the chassis has a lifting capability of 1,250 kg (2,755 lb) at 3.96 m (13 ft) and enables the vehicle to handle the Chieftain MBT engine, but not that of the Challenger, which greatly reduces the value of the vehicle to a modern battle group that may well include a squadron of Challenger tanks.

The FV 438 Anti-Tank Guided Weapons Vehicle was the first of its type to enter service with the British Army, and is only now being replaced by the 'Striker'. The British Aerospace Dynamics Swingfire missile, which is mounted in two launcher bins on the top of the rear of the hull, has a stated range of 4,000 m (13,120 ft), although in reality it would be hard pressed to make contact with a moving target at ranges in excess of 1,500 m (4,920 ft). The missile may be fired and controlled from the vehicle or remotely from a controller positioned up to 100 m (328 ft) from the vehicle

In the early 1960s it was proposed to introduce an FV 432 chassis with a Fox armoured car turret armed with a 30-mm RARDEN cannon. Although a few initial models were built and issued to Berlin Brigade, the project was scrapped at an early stage. However, many models were retro-fitted with the privately-built Peak Engineering Lightweight turret armed with a single 7.62-mm GPMG. Capable of an elevation of +50 degrees, depression of −15 degrees and 360-degree turret traverse, the system in many respects suffered from being ahead of its time. Too large to be a successful reconnaissance vehicle and appearing before the concept of the Infantry Fighting Vehicle was accepted within NATO, the design was never fully accepted and not adopted for general production.

The basic FV 432 has been used for many years by the Royal Engineers as a tow vehicle for the EMI Ranger anti-personnel mine-laying system, and more recently as an integral part of the Giant Viper mine-clearing complex. It is used by the Royal Army Medical Corps as a field ambulance, by the Royal Artillery as a mortar-locating vehicle for its Cymbeline radar, by the Royal Signals for communications and by numerous field units as a command vehicle.

The FV 432 series is now out of date and is being replaced by the much improved CVR(T) series powered by the excellent Jaguar 4.2-litre engine. These are being used for specialised tasks, while front-line mechanised and armoured units have taken the vastly more capable Warrior Mechanised Infantry Combat Vehicle into service. Twice the size of the FV 432, the Warrior is faster, more mobile, more heavily armed and much more heavily protected. It doesn't carry any more troops, however, and is considerably more expensive, and costs a lot more to run. Since the Army now has to fight for funding, it is likely that the FV 432 family will remain in service, often in the very front line, for many years to come.

Designated FV 434, the Maintenance Carrier is primarily used to change the powerpacks of Challenger Main Battle Tanks. Under the tarpaulin is a large crane able to lift over 3 tonnes (2.9 tons).

MECHANISED COMBAT

TAKING THE HIGH GROUND

Your armoured personnel carrier is bullet-proof and protects you from shell splinters, but an anti-tank missile will smash through its thin armour with horrific consequences. When on the move you must scan the ground ahead for possible killing grounds where enemy anti-tank teams are lying in wait and use the principles of fire and movement, one APC covering another.

The US Army's Field Manual FM 7-7 gives three standard strategies for APC movement. The one you select depends on the strength of your chances of making contact with the enemy.

Although this 'contact status' determines which of the three movement strategies will be used, you as platoon commander must consider the terrain and the job to be done when deciding which of the five movement formations the mounted infantry unit will adopt.

In theory, any one of them can be used in any of the three contact statuses, though in Conditions Two and Three this can be done only by splitting the platoon into two parts or by joining up with another of the Company's platoons.

1 Travelling: single-unit movement

When contact with the enemy is thought to be unlikely, your APC formation moves as a single unit, without splitting up into two elements that protect and cover each other. Because speed across the ground and control are the two most important factors, the column movement formation is used most often.

The unit is not likely to have to go into action, and it's important to keep moving as fast as possible, so the platoon commander will generally take the point position. You will use hand and arm signals to indicate the direction that the unit is to take, and also to signal changes of formation.

The APC's rear door is left open, and one member of the squad is detailed to maintain a watch on the vehicle following, reporting to you if he loses contact.

Travelling in this way, each of the platoon's vehicles is responsible for observation and first-line security in

Because of the range and power of modern weapons, a moving unit needs to have a scout element well ahead of the main body of vehicles to detect the enemy before the whole unit is in range of their weapons. Here a squad of the US 11th Armored Cavalry scans the ground ahead in Vietnam.

3 MOVEMENT STRATEGIES

1. **Travelling**
Used when contact with the enemy is unlikely.

2. **Travelling Overwatch**
Used when there is a possibility of contact with the enemy.

3. **Bounding Overwatch**
Used when contact with the enemy is expected or likely.

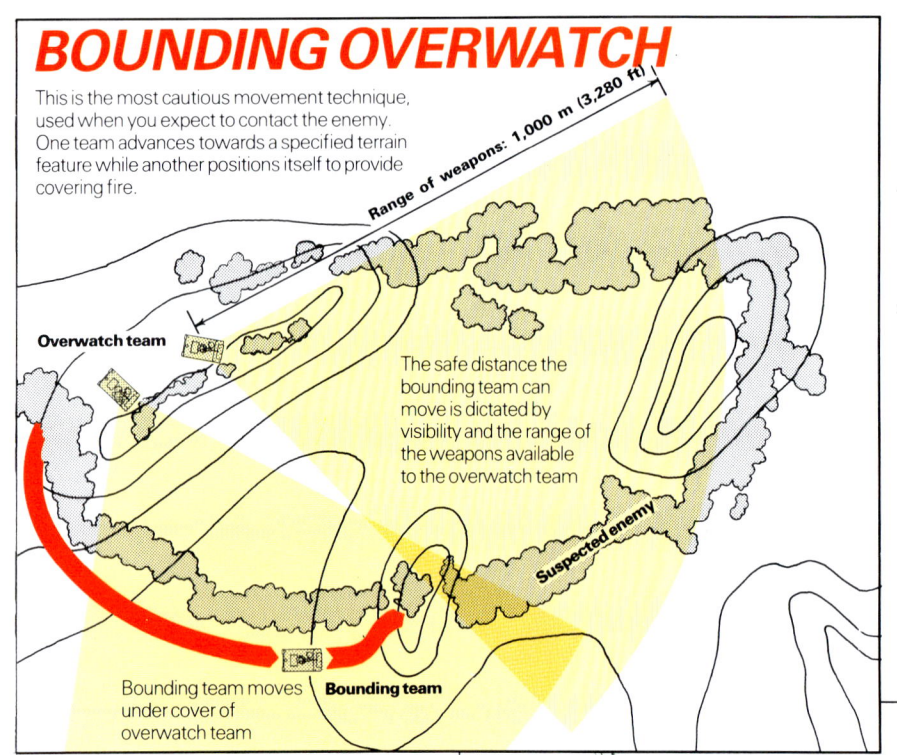

MECHANISED COMBAT

BOUNDING OVERWATCH

This is the most cautious movement technique, used when you expect to contact the enemy. One team advances towards a specified terrain feature while another positions itself to provide covering fire.

Range of weapons: 1,000 m (3,280 ft)

Overwatch team

The safe distance the bounding team can move is dictated by visibility and the range of the weapons available to the overwatch team

Suspected enemy

Bounding team moves under cover of overwatch team

Bounding team

one sector. The lead vehicle looks out ahead; the second vehicle, which is staggered to the right in column formation, to the right; the third to the left; and the fourth to the rear.

Sometimes it becomes necessary to travel with the troops dismounted from the vehicles, though this is unusual because contact with the enemy is unlikely in Condition One. If it does become necessary, you will generally dismount with your men while the platoon sergeant, your second-in-command, stays with the vehicles.

2 Travelling Overwatch: two-part movement

The second stage of readiness is called Travelling Overwatch. This technique splits the travelling force into two parts, a small spearhead group and a larger overwatch force, and requires them both to position themselves so that the larger force can cover and protect the other all the time.

The lie of the land, together with whatever information about enemy positions and strengths that's available, will decide which movement formation your unit will take up. By definition, it won't be the line formation, but any one of the other four may be used.

Moving into country that may conceal enemy forces is part normal advance and part patrol activity. It's important to keep the forward movement going, but at the same time you

Right: Hummer light vehicles armed with TOW anti-tank missiles advance using the overwatch technique in Saudi Arabia.

must present the enemy with as small a target as possible – if he can be persuaded to attack a single vehicle, he gives away his position without being able to do much good for himself in the process.

Because you must be in a position to control all four of your vehicles, you drop back to number two position in the movement formation in Travelling Overwatch, and send the lead vehicle out 400 to 600 m (1,312 to 1,970 ft) ahead, staying in visual and radio contact all the time.

Because its armour will withstand anything less than an anti-tank guided missile, the APC is well

Below: Distance between vehicles varies according to the terrain. In this Vietnamese jungle, the APCs have to keep very close indeed to be able to see each other and thus provide covering fire.

MECHANISED COMBAT

HULL DOWN TO ENEMY FIRE

Occupying a hull-down position simply means positioning the vehicle so that its hull is behind cover and cannot be hit by enemy weapons. However, the vehicle's own weapons are above the cover and able to fire at the enemy.

DAY SIGNALS

Arm and hand signals are the basic way of communicating within squads and platoons in conditions of good visibility. Because of the dangers of misunderstood signals it is important that everyone practises these signal techniques regularly. The bottom row of signals is performed by the vehicle's crew members.

I am ready — I do not understand — Assemble
Disregard previous command — Enemy in sight — Attention
Commence firing — Cease firing — Cover our move
Move out — Form line — Enemy in sight

Above: US Army arm and hand signals are used when radio silence has been imposed. Since radio traffic is so easy to detect, visual communication remains extremely important on the modern battlefield.

suited to this decoy role. If it does succeed in drawing enemy fire, the heavy .50 calibre machine-guns mounted on the platoon's vehicles will stand a very good chance of winning a fire-fight even at ranges of up to 1,000 m (3,280 ft), and the distance involved will give them every chance to reform into the most effective grouping possible in order to mount an assault.

Moving while dismounted

Even though the object of the exercise is to move forward as fast as possible, it may sometimes be necessary to move using the Travelling Overwatch technique with the platoon dismounted from the vehicles, especially if you suspect that the enemy forces may have their anti-armour specialists deployed.

In this case the lead section will take the place of the lead vehicle, and will stay in closer contact with the rest of the platoon – perhaps 100 metres in front. The vehicles must keep to positions where they can cover both the lead section and the rest of the dismounted element.

Remember, both these techniques – Travelling and Travelling Overwatch – have the same objective: to advance on an objective as fast as possible. Separating the men from their vehicles takes away the speed advantage that is the main part of the difference between Mounted Infantry and ordinary foot-soldiers. Keep the men in the carriers. Dismount only when it's absolutely necessary.

3 Bounding Overwatch: fire and move

The third movement technique, called Bounding Overwatch, is used when contact with the enemy is ex-

STREET FIGHT

In built-up areas the dismount teams lead the way in a modified column formation, clearing the buildings as they go. As the column moves under covering fire from the APCs, each team makes sure there are no enemy in the buildings on its side of the street and keeps the upper floors of the buildings across the street under observation.

MECHANISED COMBAT

pected. The attacking force is split into two equal parts, the bounding force and the overwatch force, and are used in a way very similar to the traditional infantryman's fire-and-move tactics.

Bounding Overwatch is the most deliberate and cautious of the three movement techniques. While the other two assume that the enemy may be about, and arrange the unit to counter any move he may make, Bounding Overwatch assumes that the enemy is definitely there waiting to attack. The other two put speed first, but the Bounding Overwatch is designed to make the operation as safe as possible for the troops and vehicles taking part.

Approaching the enemy

The overwatch force covers the bounding force from a static position that offers a good field of fire against possible enemy operations. How far the bounding force will go is decided in advance. The sort of things that you will look for when selecting a target site depends on which of two types of movement you use at the time. These could be:

1 Successive advance, where the overwatch force moves up to the positions that the bounding force has just established, takes them over and covers the next movement of the bounding force.

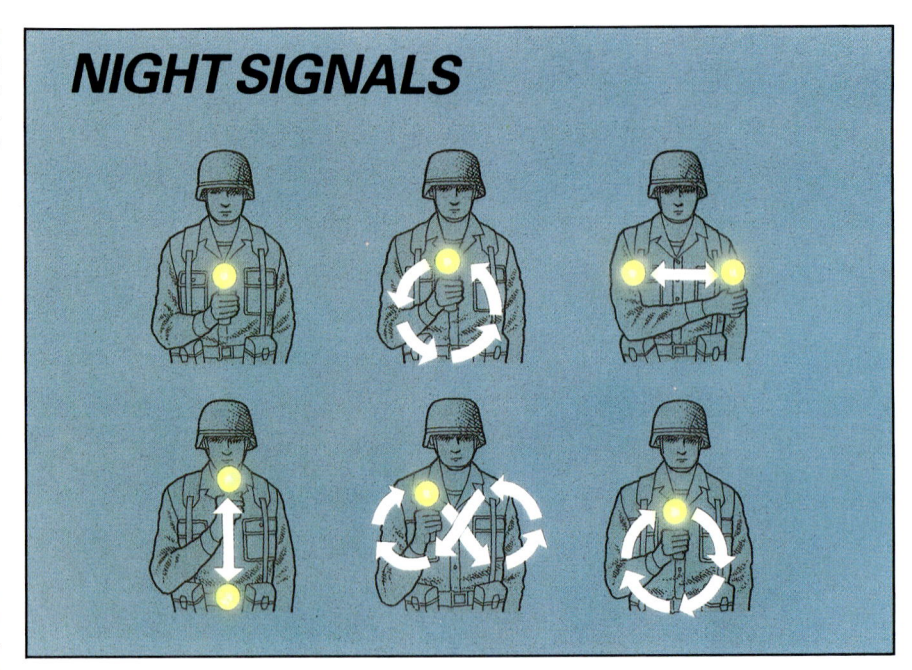

2 Alternate advance, where the overwatch force moves forward through and past the area where the bounding force has come to a halt, and takes on the job of the bounding force itself.

The length of each bound is limited by the effective range of the weapons available, by the fields of fire that the overwatch force can control, and by other, more artificial factors such as visibility in bad weather.

APCs enable the infantry to keep up with tanks, but sometimes the tanks have to lend a hand: here an M551 Sheridan tows an M113 through deep mud in Vietnam.

MECHANISED COMBAT

HITTING THE BEACH
WITH THE AAV7

'Hitting beaches' — that is what most people's perception would be of the role of marine infantry in modern warfare: to be able to land directly from some form of seaborne assault transport onto a defended shoreline, to survive the critical period of disadvantage struggling from one environment to another, then to establish and hold a beach-head.
Such techniques of 'amphibious' warfare were developed to a high degree during World War II.

The success of these operations was greatly assisted by a new kind of fighting vehicle; classified by the US Army and Marine Corps (USMC) as the Landing Vehicle Tracked (LVT), it was a cross between a landing craft and a tank. Its watertight hull meant it could float and be steered like a boat, while its tracks meant it could move freely on land.

After World War II few members of the British armed forces ever expected they would have to make an amphi-

Using their AAV7 amphibious assault vehicles as APCs, the US Marines formed the east flank of the coalition attack on Iraq's invasion force in Kuwait.

bious assault over open beaches, but that is precisely what happened in 1982 during Operation Corporate, the successful British operation to re-take the Falkland Islands from the Argentine invaders. In fact, British infantry assaulted from open landing craft without any specialised armoured vehicles to support them.

LVTP7
Ironically it was the Argentine Marines who had such vehicles. They were equipped with the most modern US LVT in service, the LVTP7. The Argentines did not employ them in the initial assault on the tiny British garrison but landed them later by

Originally known as the LVTP7 (or Landing Vehicle, Tracked, Personnel, Mk 7), the AAV7 has been the US Marine Corps' main seaborne assault vehicle since the 1970s.

MECHANISED COMBAT

The AAV7 has been well tested in arid conditions; the Marine Corps Air-Ground Combat Center is at Twentynine Palms in California's Mojave desert.

Inside the AAV7

The AAV7 will soldier on in with the US Marine amphibious assault battalions until the late 1990s, when a replacement advanced amphibious assault vehicle should begin to enter service. There are no vehicles comparable to the AAV7 currently in service anywhere in the world: its combination of load capacity, endurance, amphibious ability and fighting power make it unique.

ship, using them as armoured personnel carriers in the Port Stanley area.

In the early 1960s the standard US Marine Corps LVTP ('P' for personnel) was the LVTP5A1, a mid-1950s vintage design with short range on land and in water. In early 1966 development of a new vehicle began and the first prototype trials (the experimental vehicles being dubbed the LVTPX12) were completed in September 1969. FMC won the production contract (for 942 vehicles), and the first LVTP Model 7 was delivered to the US Marines in 1972.

The vehicle was a considerable improvement over the LVTP5: it was faster on road (64 km/h [40 mph] as compared with 48 km/h [30 mph]), had a greater road range (482 km [299.5 miles]), and slashed the number of maintenance hours required per hundred hours of operation from 22 to six. Track life was also trebled to 600 hours.

Plenty of room

Redesignated in 1985 as the Amphibious Assault Vehicle 7 or AAV7, the amphibian is front-engined with a spacious troop compartment in the rear of the boat-shaped hull accommodating 25 fully-equipped marines on three bench seats, who can enter and leave via a power-operated ramp at the rear. They can also exit through spring-balanced roof hatches.

The vehicle's all-welded aluminium hull gives protection against shell splinters, small arms fire and flash burns, but would be vulnerable to high-velocity shells or anti-tank missiles. In its role as a sea to shore amphibious assault craft this level of protection is judged adequate.

The crew of three comprises the driver, who sits at the front of the hull on the left side; the commander, who sits directly behind the driver; and a gunner, who operates the vehicle's only armament, a 12.7-mm M85 machine-gun mounted in an electro-

Machine-gun
The AAV7's M58 .50-calibre machine-gun is mounted in a small turret behind the engine. One thousand rounds are carried, and they can be fired at a high rate of 1,050 rounds per minute or at a low rate of 450 rounds per minute.

Fan and radiator
Air for the AAV7's engine is drawn in and discharged through grilles in the roof, which are sealed while the vehicle is in water.

Powerpack
The AAV7 was originally powered by an eight-cylinder Detroit Diesel, developing some 400 hp. Later models have been fitted with a Cummins VT 400 Diesel instead. Either engine can be removed in about 45 minutes.

Frontal armour
The aluminium armour at the front of the AAV7 is 45 mm (1.8 in) thick, and gives protection against small-arms fire from weapons of up to 14.5-mm calibre.

The AAV7 is a large vehicle, able to carry far more troops than comparable land-based infantry fighting vehicles but having far less armour.

MECHANISED COMBAT

Roof hatches
The AAV7 has three torsion-spring-balanced roof hatches which are used to load and unload the vehicle when it is in water.

Power ramp
The power-operated ramp in the rear of the vehicle has a small door built-in to allow access in the event of mechanical failure.

Driver
The driver's station is immediately in front of the commander. It can be fitted with an M24 infrared periscope for use at night.

Commander
The vehicle commander's station is equipped with vision blocks and an M17C periscope, which is extendable to give a view over the driver's hatch.

Side armour
The hull top, sides and bottom of the AAV7 are protected by 30 mm (1.2 in) of aluminium armour, giving protection against small-arms fire and shell splinters.

Below: Marines are expected to be able to fight wherever their country sends them, which means that equipment like the AAV7 must be expected to work from the jungle to the Arctic.

hydraulically powered turret offset to the vehicle's left.

No ports are provided for small arms and personnel cannot 'fight' from inside a closed-up vehicle.

The AAV7 is fully amphibious without preparation, propulsion in water being provided by two water jets in the hull rear. At the rear of each jet is a hinged rudder which can be fully deflected to put the vehicle into reverse. Maximum speed forwards in water is 13.5 km/h (8.4 mph) and backwards 7.2 km/h (4.5 mph), and the hull has been carefully designed to be as stable as possible, even in high sea states.

Between 1982 and 1986 the USMC's entire fleet of amphibious vehicles was reworked to AAV7A1 standard, with an emphasis on improving the

MECHANISED COMBAT

vehicle's durability and ease of maintenance, which are always a problem in the especially demanding environment of amphibious operations.

Possible improvements

The USMC began looking at possible successors for the AAV7 in the mid-1970s, and much research effort was spent on high-speed armoured hovercraft and other exotic solutions, but the LVTX (for Experimental) programme was terminated in 1985 on the grounds of cost and the fact that the AAV7A1 could be product-improved to meet the requirements of the 1990s and beyond.

The major problem is affording the vehicle some fighting ability of its own, upgrading it from a swimming armoured bus to a swimming armoured fighting machine. To this end the USMC has considered several additional improvements for the vehicle, including adding a retractable bow plate to improve handling in water, add-on appliqué armour for forward deployed vehicles, adding a 40-mm grenade thrower in the turret and, significantly, adding a universal weapon mount for TOW or Dragon anti-tank missiles.

In late 1986 the US Navy Sea Systems Command ordered 240 'Upgunned Weapons Stations' from the Cadillac Gage company to upgrade AAFV7A1s with a new tactical turret combining a 40-mm grenade launcher and 12.7-mm M2 machine-gun.

The LVTP7A1 (or AAV7A1, as it is more correctly called) is an early 1970s design which appears to be capable of a good degree of 'stretch' and seems set to be standard equipment for the US Marine Corps into the next century. In that service they are issued to Assault Amphibian Battalions which consist of a Headquarters and Service Company and four Assault Amphibian Companies.

The HQ and Service Company has 15 AAV7A1s, three AAVC7A1s (command vehicle conversions), and one AAVR7A1 (recovery vehicle conversion). The Assault Amphibian Company has four platoons of AAV7A1s, each platoon having 10 vehicles, which, with the company HQ section, makes a total of 207 vehicles. Other marine corps which operate the vehicle include those of Argentina, already mentioned, Thailand, Spain, Italy, South Korea and Venezuela.

AAV7s went into large-scale action in the Gulf in 1991. A two-division Marine Expeditionary Force was heavily involved in the war to eject the Iraqi army from Kuwait. Although there were plenty of AAV7s afloat in the coalition's amphibious groups, it was as a conventional land-bound APC that it saw most action.

An AAV7 rumbles onto a US Navy assault ship after a three-kilometre swim from the beach. AAV7s take 15 minutes to travel that distance in water.

AAV7s move down a South Carolina track during a mock assault, part of a major multi-unit exercise the US Marine Corps runs nearly every year.

MECHANISED COMBAT

THE COVERING FORCE

THE COVERING FORCE MISSION

The covering force, composed of approximately one division, will make first contact with the enemy and will fight an aggressive delaying action designed to achieve three aims:

1. To cause maximum damage and disruption to the leading enemy elements and formations.
2. To delay the enemy for a specific period while the main defensive force deploy and dig in the main defensive position.
3. To identify and assess the strength and direction of the enemy advance and identify those areas targeted for the main enemy thrust.

Covering troops fight a delaying battle forward of the main position, to delay the enemy and discover the main axes of his advance. They are normally divided – into a 'screen', based on armoured reconnaissance regiments, and 'delaying forces' and reserves, based on armoured battlegroups.

The screen will consist of squadrons of armoured reconnaissance troops mounted in Scimitar, Striker and Spartan reconnaissance vehicles. These are all part of the Combat Vehicle Reconnaissance (Tracked) or CVR(T) family.

A medium recce squadron in BAOR contains three troops of four Scimitars, a troop of four Strikers and a troop of four Spartans. Scimitar is

The eyes of the Battle Group, the Medium Recce Squadron, is equipped with Scimitar, Striker and Spartan. Scimitar is designed to be deployed well forward to recce the enemy and then break clean to report back. Russian recce, equipped with Main Battle Tanks, would make short work of the Scimitar, and the Scimitar's 30-mm cannon would hardly give an MBT crew a headache.

MECHANISED COMBAT

Above: M1 Abrams tanks in the Gulf War. The M1's cross-country performance is so good that it is actually faster than many specialist reconnaissance vehicles despite their theoretically higher speed.

armed with a 30-mm RARDEN cannon and a 7.62-mm machine-gun; Striker is equipped with the Swingfire ATGW. Spartan carries a small section of assault troopers armed as infantrymen.

These vehicles work in pairs and operate as stealthily as possible so as not to be detected. Their task is to identify enemy axes of advance and they must, at all costs, avoid an entanglement with the enemy. They are not equipped to get involved in a fight, except perhaps against recce vehicles such as the Soviet BRDM. They engage enemy vehicles only in self-defence. The navigation, camouflage and driving skills of their crews must be superb, as they will often be outflanked and may have to extricate themselves through enemy forward elements.

A delaying force (or a reserve) should contain a high proportion of tanks and anti-tank guided weapons (ATGW), and should be supported by reconnaissance helicopters and engineers.

These mean they can deal with the enemy at arm's length; it is dangerous for any part of a covering force to become entangled at close quarters with the enemy. Their task is to cause delay by knocking out enemy forces at relatively long range. Guided weapons can fire out to 4,000 m (13,120 ft) (Swingfire) or 2,000 m (6,560 ft) (MILAN); Challenger tanks can engage targets at over 2,000 m (6,560 ft).

Battlegroups in the delaying force will be expected to occupy a series of delaying positions. From each one, they may break contact with the enemy and leapfrog to a new position behind other delaying forces. Or they could withdraw in contact with the enemy to a new position in depth. This latter alternative is sometimes known as a fighting withdrawal. In either case it will be a battle of movement in which the enemy is delayed by fire. The action is really a mixture of a fighting withdrawal and hasty or improvised defence.

What to do

When carrying out such a mission, you should remember certain rules: first, even if you cannot defend everywhere, you must at least be able to *see* everywhere. In other words, you must maintain an unbroken line of surveillance across the battlegroup front.

Second, your series of delaying positions should – as with any defensive position – take full advantage of any natural obstacles, such as a river line or ridge line, augmented wher-

Air defence of US Army recce units is being boosted by extra Stinger missiles, but hand-held SAM teams do not tend to survive for long in combat. US Army research gives them a life expectancy of three minutes after firing.

ever possible by artifical obstacles.

Third, you must at all costs avoid the 'thin red line' mentality. Concentrate your resources on the most probable enemy axes of advance. You then release sufficient troops to provide depth and mobile reserves. You will need these to stop the enemy penetrating your position, outflanking you, or threatening your withdrawal routes with heliborne troops.

Fourth, make maximum use of long-range anti-tank engagements, particularly with ATGW, to defeat enemy reconnaissance and force him into deploying to attack you. Just as he is about to attack, you withdraw.

Finally, never forget the air threat, particularly from armed helicopters and troop-carrying helicopters. You will need to have thought out an air defence plan using your Rapier and Blowpipe missiles (if you are lucky enough to have both). Site them to give your armour the best protection from the threat of air attack.

MECHANISED COMBAT

The covering force will be supported by tank-busting helicopters and close air support from Harriers and A-10s. The SNEB rocket pods seen here are now only used in Belize.

Above: The Striker provides the anti-tank teeth of the Medium Recce Squadron and will destroy most tanks out to 4,000 m (13,120 ft). Each vehicle carries 10 rounds.

Below: The Kurassier Austrian light tank carries a 105-mm light gun capable of knocking out an MBT. However, it does not have the manoeuvrability or speed of the lighter dedicated recce vehicles, and would therefore be more likely to get into a fight in the first place.

Trip up the enemy

There are certain factors that affect your planning and deserve special attention from covering troops. The first is the 'obstacle plan'. If you have sufficient warning time, there is an enormous amount your engineers can do to slow the enemy down while, at the same time, making your withdrawal easier.

Minefields, anti-tank ditches, demolitions, tree-felling and much else can be set up if you have enough time. But – this is terribly important – your obstacle plan must not be so comprehensive that it interferes with your withdrawal.

Below: The covering force must not get involved in protracted battles with the enemy, as this will provide them with the opportunity to make full and accurate use of artillery. This 130-mm M46 field gun is seen here in Iraqi service against Iran.

The second important factor to consider in planning your covering force battle is one great advantage you have over the enemy – your knowledge of the ground. You should have reconnoitred every track and road, every withdrawal route and alternative route, every field of fire, alternative firing position, covered approach and so on. The enemy may outnumber you, but you are fighting on your own ground. Make the most of it.

The battle will be complicated, confusing and fast-moving. Careful route planning, movement control and co-ordination is vitally important. Pay particular attention to the formations on either flank.

It is very important that your plans

MECHANISED COMBAT

MLRS has considerably enhanced NATO's capability to fight the covering force battle; MLRS can put vast quantities of HE on a piece of real estate very quickly. This firepower could be employed very effectively on counter battery fire tasks.

are co-ordinated and synchronised with flanking units. However well you are fighting your battle, the whole pack of cards is likely to fall down if the covering force on your left has decided to withdraw at twice your rate.

Surprise

Last, but almost certainly most important, you must surprise and deceive the enemy. If he is able to anticipate your withdrawal, he may be able to turn it into a rout. You need effective camouflage and concealment measures, and absolute radio security, to conceal not only your locations but also your overall plan and intentions. There is tremendous scope for low level deception measures.

This will be the first time the enemy has crossed swords with you, and he will not know quite what to expect. Ideally you would prepare complete dummy positions and deploy dummy equipment and decoys. But, if time does not permit such detailed preparations, simple deceptive ploys can be very effective. You only have to create track marks leading into cover to give the impression to aerial reconnaissance that tanks are lurking there. The odd camouflage net over some angular construction may well give the impression from a distance that it is a camouflaged vehicle.

Why withdraw?

Follow these principles and bear these planning factors in mind, and you will have the essentials of the covering force battle. Above all, it is crucial that the commander maintains the balance of his forces throughout the withdrawal, to be able to react quickly and effectively to unexpected developments.

A withdrawal is not the easiest battle to fight – and the covering force battle is essentially a withdrawal – as it places great strain on the confidence and morale of all ranks. It needs decisive and confident leadership and a cool head. But a successful covering force battle can cause serious loses to the enemy, provide valuable time for the main defensive forces to prepare their positions and logistic supplies for the forthcoming battle, and give important information on enemy intentions, tactics and capabilities.

Above: The covering force must be able to break clean, which means maintaining their mobility on the battlefield while slowing the enemy up with obstacles. Combat engineers would be responsible for these tasks.

Below: A Wombat 120-mm anti-tank weapon team, having expended all their ammunition and taken casualties, prepare to withdraw as part of a covering force exercise. Present doctrine relies on there being enough of this force left to replenish and form the mobile reserve.

MECHANISED COMBAT

RACING TO WAR
WITH THE
WARRIOR

It is twice as heavy as the FV 432 it replaces and it carries fewer men, but the Warrior mechanised combat vehicle has greatly increased the power of Britain's front-line infantry units.

Fast, heavily armed and superbly agile, the British Army's Warrior mechanised combat vehicle will take you into battle safely, at speed, and in style. It represents a new philosophy for British combat infantry. The Warrior replaces the FV 432 APC that has served the Army since the early 1960s and followed the successful design pioneered by the US M113 APC. Armoured infantry philosophy at the time demanded a vehicle that could get a section of infantry from A to B on the battlefield in comparative safety. In other words it was an armoured taxi designed to deliver you somewhere short of your objective. Then, you would debus and assault on foot. The FV432 was not designed to fight from.

Pace and punch

British thinking has, over the years, changed considerably. The result is the MCV (Mechanised Combat Vehicle) 80 Warrior. This vehicle, as its name implies, is essentially a *fighting* vehicle. It has the power and mobility to keep up with the Challenger tank and is armed with a 30-mm RARDEN cannon and a Hughes Chain Gun, although unlike many contemporaries, it does not have firing ports. The British Army doubts their value in combat, so the section will still have to dismount short of the objective and continue on foot.

The Warrior's unexpected combat debut came in the Gulf War, when together with the Challenger tank it equipped the British part of the spectacular coalition drive around the Iraqi forces illegally occupying Kuwait.

MECHANISED COMBAT

However, Warrior will provide powerful covering fire when you go into assault. And in defence, your Warrior can be dug in, hull-down, behind your position so that it can add its weight of fire to the section's defence.

Go anywhere

Warrior weighs 25.4 tonnes (25 tons), very nearly twice the weight of the old FV 432. It is also longer, wider and higher than the 432. But it is in the performance that you will notice the difference. Whereas the old 432 used to lumber along at a theoretical top speed of 50 km/h (31.25 mph) (actually it was nearer 40 km/h [25 mph] because the vehicles were getting so old), Warrior's top road speed is over 80 km/h (50 mph). But its agility, mobility and acceleration are truly impressive: it can reach 48 km/h (30 mph) in 18 seconds.

The combination of aluminium alloy armour, a powerful engine and a remarkable suspension system means that you can move impressively fast between cover on the battlefield. This means that you and the rest of your section will be exposed to enemy fire for shorter periods of time.

This is terribly important. Modern anti-tank guided missiles fly quite fast out to their maximum range (MILAN, for instance, takes 13 seconds to reach 2,000 m [6,560 ft]). Thus, if you are 1,500 m (4,920 ft) from an enemy ATGW system, you probably have about 10 seconds (including acquisition time) to motor from one bit of cover to the next. In Warrior, depending on terrain, that might be possible. In the 432 it almost certainly is not.

And in Warrior you can shoot back. It is armed with a 30-mm RARDEN cannon, which is capable of firing

Warrior could have been made as low as the Soviet BMP series, but the decision was made to increase the headroom in the troop compartment. This makes life much more comfortable for the men in the back.

Inside the Warrior

Warrior will change British mechanised infantry out of all recognition. Now called Armoured Infantry, battalions will need increased manpower with an additional junior NCO in every Warrior. He will command the vehicle when the section commander and 2-i-c dismount with the infantry fireteams.

Commander

7.62-mm Chain Gun
Electrically driven, the Chain Gun ejects a hail of cases out of the turret and is a great improvement over most turret-mounted machine-guns because it does not fill the turret with fumes.

RARDEN 30-mm cannon

Smoke rocket dischargers

Driver
Driving the Warrior is enormous fun; its speed and acceleration are a delight. On German autobahns and British motorways, Warriors have been found exceeding the speed limit. Full marks to the vehicle, but not to the drivers concerned!

Aluminium hull armour
This is used instead of steel to reduce the weight of the vehicle.

APDS and HE rounds out to 2,000 m (6,560 ft). Two hundred and twenty-five rounds are carried inside the turret. The 7.62-mm Hughes Chain Gun, mounted co-axially in the turret, can shoot out to 1,100 m (3,610 ft).

Answer back

Despite this considerable firepower, you should not use Warrior as a tank. It has nothing like the same sort of protection as Challenger – no APC or MCV has. The RARDEN is designed to take out enemy APCs, and the Hughes Chain Gun is designed to support you when you are dismounted from the vehicle. If the gunner in a Warrior was foolish enough to take on a tank, there is no doubt who would come off worst.

The vehicle does have some anti-

MECHANISED COMBAT

Day/night sights
420,000-worth of image intensifier gives the commander and gunner ×8 magnification and night vision capability.

Steel turret armour
The most likely area to be struck by enemy rounds, the turret is protected by steel armour.

Gunner

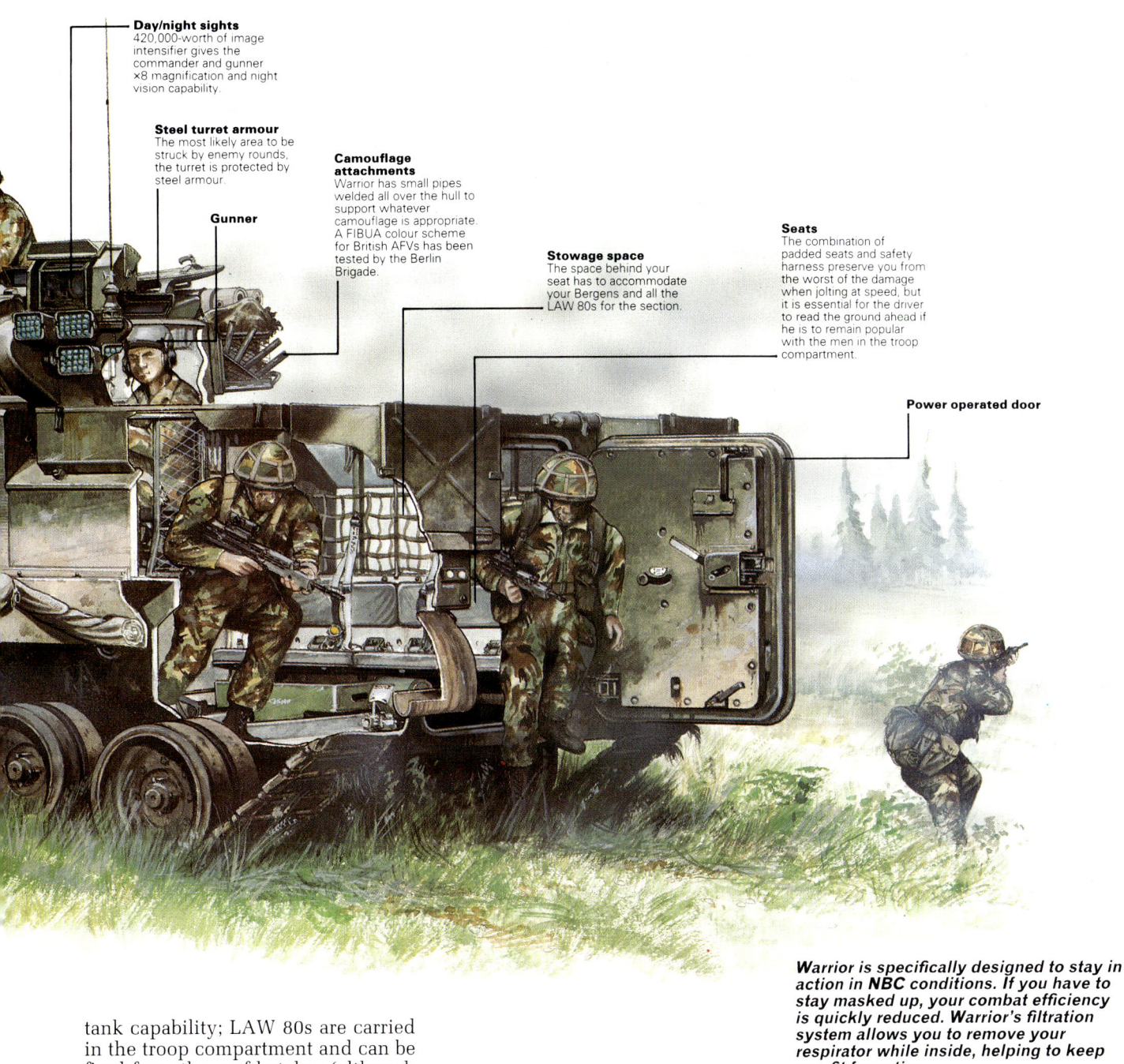

Camouflage attachments
Warrior has small pipes welded all over the hull to support whatever camouflage is appropriate. A FIBUA colour scheme for British AFVs has been tested by the Berlin Brigade.

Stowage space
The space behind your seat has to accommodate your Bergens and all the LAW 80s for the section.

Seats
The combination of padded seats and safety harness preserve you from the worst of the damage when jolting at speed, but it is essential for the driver to read the ground ahead if he is to remain popular with the men in the troop compartment.

Power operated door

Warrior is specifically designed to stay in action in NBC conditions. If you have to stay masked up, your combat efficiency is quickly reduced. Warrior's filtration system allows you to remove your respirator while inside, helping to keep you fit for action.

tank capability; LAW 80s are carried in the troop compartment and can be fired from the roof hatches (although normally the section will be dismounted when firing them). They can even be fired from the turret by the commander or gunner. Firing from the vehicle could be a useful tactic in close country or built-up areas.

Warrior is also equipped with multi-barrel smoke dischargers, which are mounted either side of the turret and fire forwards. They discharge a pattern of smoke grenades to create an instant smoke screen between you and the enemy. This is often useful if you are under fire and need to extricate yourself quickly. The turret is equipped with a Raven ×8 day sight and fully integrated image intensifier (II) night sight, so that the vehicle can operate 24 hours a day.

Warrior is designed to carry a total

MECHANISED COMBAT

of 10 men, including a vehicle commander and driver. Because Warrior is designed to support the section when it dismounts, the vehicle commander remains with the vehicle, acquiring targets and loading the RARDEN cannon when the section dismounts. Two fireteams, one of four men and one of three, will debus, both armed with an LSW. The vehicle itself will act as a third fireteam. The section commander will lead one team, the section second-in-command will command the other, and an additional JNCO will command the vehicle.

The crew compartment is inevitably cramped but has been carefully designed to give you reasonable headroom. The latest suspension system is excellent but you still need the individual harness during high-speed cross-country driving. There is the usual large container for boiling hot water electrically, which means you can still indulge in the all-important 'brews' to keep body and soul together! But, even more important, the vehicle is fitted with a highly efficient air filtration system.

Most new armoured vehicles spawn a whole range of variants and become, in time, a 'family' of vehicles. Warrior has been designed so that it can easily be adapted for any number of roles. Thirteen variants are planned, including two versions of an APC, a command vehicle, a recovery vehicle, a combat repair vehicle, mortar, ATGW, recce and anti-aircraft vehicles, rocket-launcher and load-carrying vehicles, and even a 30-tonne light battle tank mounting a 105-mm gun. Some of these variants exist and will be brought into service with the British Army; others remain on the drawing board. The Warrior project is an exciting one, with a great future. The main job of Warrior, though, is to carry the infantryman into battle and support him when he gets there.

The Warrior's combat debut came in 1991, when the infantry battalions of Britain's 1st Armoured Division went into action against Saddam Hussein's Iraqis. The desert sands and heat caused few mechanical problems, although as with all modern infantry combat vehicles the Warrior's crews complained of being somewhat cramped. The Warrior's mobility enabled it to keep up with the tanks, but it did create another problem. Many infantry support elements, such as ambulances and mortars, were mounted in old FV 432s, and they just could not keep up with the fast-moving armoured columns as they pushed into Kuwait and southern Iraq.

One problem the British Army encountered in the Gulf was that vehicles based on the old FV 432 – command vehicles, ambulances and the like – had great difficulty in keeping up with the far more mobile units mounted in Warriors.

The British Army has taken delivery of a number of Warrior variants. This is the repair and recovery vehicle. The hydraulically operated crane mounted on the left side of the vehicle has 6,500 kg (14,330 lb) capacity and can lift out a complete Challenger tank powerpack.

MECHANISED COMBAT

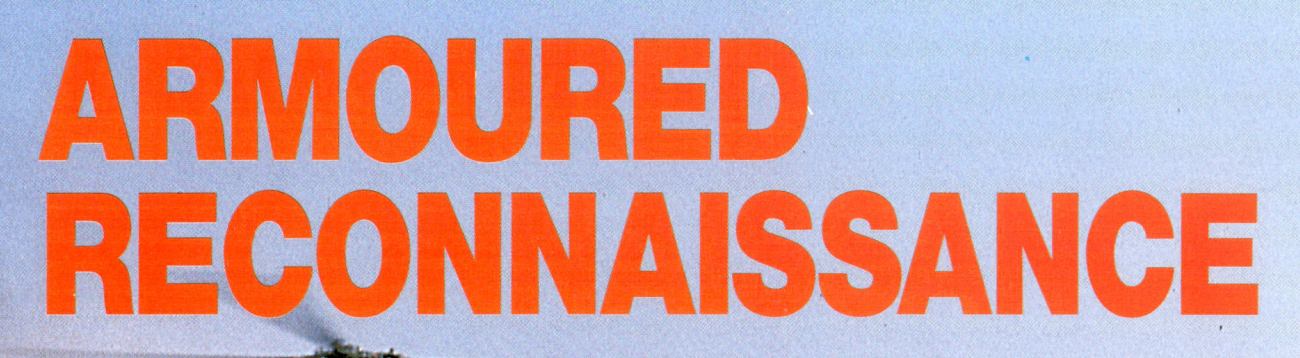

ARMOURED RECONNAISSANCE

The eyes and ears of the armoured battlegroup: Gazelle helicopters and CVR(T)s work together, locating enemy forces well away from the main body. The information they supply to the battlegroup headquarters will allow the full strength of the armoured forces to be deployed in the right place.

The blinding white flash of the round hitting the target was immediately followed by a huge fireball as the ammo inside the now burning hulk went up. The frying-pan-shaped turret of the T-62 was flung into the air, landing next to the flaming hull. Obviously the ancient T-62 was no match for a Challenger and a Fin round from 2,000 m (6,560 ft).

But where there is one T-62 there are likely to be more, and it's pretty safe to bet that the tank was part of a forward screen of some kind of position. Now is the time to get on the radio to the squadron leader and tell him what is happening.

Without a doubt, the most important thing for the squadron leader is information. If there is a position ahead, he must know about it as soon as possible; where it is, what size it is, what equipment is there, possible location of supporting positions, locations of any reserve... the list is huge.

The Challenger is a mighty fine tank and well suited to many things, but trying to conduct recce by stealth in a 1,200-hp, 60-tonne (59 ton) tank is asking just a bit too much. Since yours is the first contact across the battlegroup's front, the recce troop will be heading there as fast as they can.

Scorpions and Scimitars

Recce troop are the eyes and ears of the battlegroup; eight CVR(T)s, either the 76-mm Scorpion for an armoured battlegroup, or the 30-mm Scimitar if you are working with an infantry one. Their task is to recce ahead and give just the sort of information you need for the attack.

Always operating in pairs to provide mutual support, they will conduct a probing recce to locate and identify the position. Sure enough, recce report the sighting of a platoon-sized anti-tank position covering the axis of advance.

The only real question is how big the attacking force is going to be. The rule of thumb is that defenders have a 3:1 advantage over attackers; it requires three attacking tanks to take out one defending tank.

Battlegroups are made by combining sub-units of tank squadrons and infantry companies under the command of a battlegroup HQ – either an infantry HQ or an armoured HQ. The mix of sub-units depends on the tasks anticipated, but it is uncommon to have more than five or less than three sub-units.

MECHANISED COMBAT

Recce have done their work. The position is, indeed, platoon-sized and appears to be unsupported. It has a large number of anti-tank weapons and appears to be covering a wadi – a dried river bed – which marks the axis of advance. They are well dug in.

The squadron leader will now formulate his plan. He will divide his tank force into three components. To destroy any enemy armour on the position and to put down heavy, accurate fire, one troop will be detailed off to be the fire support troop. The three remaining troops will be divided into two groups; two troops become the assault troops (one left, one right) and one troop becomes the intimate support troop.

But before all that happens, there is some real estate that needs to be sorted out. So far recce have done an excellent job in locating and identifying the enemy position. However, their work is not over. Since they are the only people who know what the land is like, they are in an ideal position to select an FUP and fire support position.

The FUP needs to be big: it will get very crowded with all the armoured vehicles in it. It must be out of sight of the enemy, who must not be able to bring direct fire in to it. It should not

The conventional M551 Sheridan light tank was retained by US cavalry battalions posted in Germany for many years. Its 152-mm gun/missile system was capable of destroying all Russian tanks, but its reliability was always a problem.

be so blindingly obvious a position that they already have it registered as an artillery position. It must be in a direct line with the objective and not too far from it – not more than 1,500 m (4,920 ft) if possible (in the desert, FUPs are likely to be much further) and, most important of all, it must be secure: there must be no enemy sitting in it when you arrive.

Fire support

The fire support position is equally as important. It should be just off to a flank – about 30 degrees off the axis of advance; any more and you are going to expose a flank. The most important thing is that the fire support troop can get to it without revealing their position and can bring down accurate fire onto the entire objective.

As soon as the recce have passed the information back, the squadron leader can make a plan and issue a warning order. The warning order will contain an outline of the plan, detail troops to tasks and give locations and likely timings. Once you have those you can swing into action. The fire support troop will start to motor off to its position, under the command of the squadron second-in-command. The remainder will begin to move round to the FUP.

A Scorpion of the 17th/21st Lancers, festooned with local camouflage, crashes across the Soltau training area in Germany. The Scorpion's 76-mm gun fires a HESH round capable of destroying enemy recce vehicles.

MECHANISED COMBAT

Above: The Russian army's recce battalions included PT-76 amphibious light tanks, capable of tackling most water obstacles.

Below: A handful of CVR(T)s were sent to the Falklands, where they demonstrated their excellent ability to cope with boggy ground.

H-hour, the exact plan and, for the infantry, dismounting instructions.

The move round to the FUP needs to be slick and precise. When you get there everybody must know exactly where to go and what they are doing. The assaulting tanks will be on both flanks, with the intimate support troop in the centre, the Warriors lined up behind them.

No sooner is the last vehicle approaching than the artillery fire plan begins. From 15 km (9.4 miles) away a regiment of M109s opens up. Twenty-four barrels of 155-mm artillery spit fire into the sky and seconds later the ground around the objective seems to boil as shell after shell explodes.

With one minute to go before H-hour, the fire support tanks roll the last few metres into position and im-

The limiting factor in an attack is the availability of artillery. If you are lucky they will not be on other tasks and will be available for the attack. Since you cannot attack without artillery fire, the time when the attack is launched, or H-hour, will be dictated by the gunners.

This time you are lucky – the guns are not on task and the attack has been given priority. Even on a relatively small position, you should have a lot of guns firing for a short period rather than a few guns popping away for an eternity.

While the squadron/company group is moving round to the FUP, the squadron leader and company commander give the final plan. Most of the orders will just confirm the warning orders, except now you have a definite

The 30-mm RARDEN cannon fitted to the Scimitars can not only destroy light armoured vehicles; in the Gulf War it proved highly capable of penetrating enemy bunkers.

MECHANISED COMBAT

Above: Identifying water crossings is a critical task for the reconnaissance forces. After over 40 years in Germany most rivers have been surveyed by engineers, but with overseas deployments you must obtain the information yourself.

Above: Urban areas present a serious problem for armoured reconnaissance units. The enemy could be concealed anywhere, but you can no longer just drive around – many parts of Europe are simply too built-up.

The BRDM-2 armoured car is one of the most commonly encountered reconnaissance vehicles in the world. It is built for reliability in the field rather than for speed or armament. Its cross-country performance is good and it is armed with a 14.5-mm heavy machine-gun.

mediately bring down pinpoint accurate fire onto the few anti-tank vehicles on the position. Using their thermal sights they penetrate the dust and smoke kicked up by the artillery, which provides no cover for the BMPs and BRDMs.

Into battle

On the very second of H-hour the entire armoured force moves, as one, over the ridge and into battle. The assault tanks fire on the move into the enemy position with both the main gun and the 7.62-mm co-axial machine-gun. As the fire plan lifts, the lead tanks are 400 m (1,310 ft) from the front of the position when the two assault troops split left and right, going round the objective firing into it.

The way is left open for the intimate support troop to lead the infantry right into the very heart of the objective. Using both the 120-mm main gun from blank range and the machine-gun, they provide overwhelming local firepower.

On the front edge the tanks and APCs stamp on the brakes. The backs of the Warriors fly open and out pour the infantry, covered by a massive weight of fire from the tanks and the Warrior's 30-mm RARDEN cannon. It is now up to the infantry.

Meanwhile the assault tanks will have gone firm in a ring around the position, sealing it in and preventing the enemy from counter-attacking. As soon as the intimate support tanks have hit the position, the fire support troop and second-in-command will join that ring of steel.

MECHANISED COMBAT

DRAGOON

In war, something that can go wrong will usually do so, and often at a very inconvenient moment. This is especially true of military vehicles, from jeeps to Main Battle Tanks. The truck containing the spare parts is usually the one that hits a mine, bogs down or gets stuck the wrong side of a blown bridge. NATO armies talk endlessly about the need to introduce compatible equipment so that allied forces can use each other's kit, but few practical steps have been taken. However, one small manufacturer has designed a family of combat vehicles to soldier on regardless.

The Dragoon is a four-wheel-drive, all-terrain vehicle currently in service with the US Army and Navy as well as certain police agencies, which use it for special operations. The secret of its success is interchangeability – 80 per cent of its parts and components are common with those of two other vehicles: the ubiquitous M113 armoured personnel carrier and the US Army's standard 5.1-tonne (5-ton). The M113 and M809 truck have been produced in such staggering numbers and are used by so many of the world's armies that the supply of spares and the availability of trained mechanics is assured almost anywhere.

The Dragoon is the progenitor of a versatile family of armoured fighting vehicles. It is available in a variety of specialist forms in addition to the basic armoured personnel carrier and armoured reconnaissance vehicle variants.

Effective APC

The Dragoon series was conceived in 1976 by the Arrowpointe Corporation to meet the security and tactical support needs of the US military. The aim was to produce an effective weapons or personnel carrier at a very competitive price. It was not intended to compete with the M113s used by the mechanised forces, but to perform the many roles for which a tracked APC was inappropriate. For example, a

The assault gun version mounts a 90-mm Cockerill gun in a two-man turret. Improvements in 90-mm ammunition, including APFSDS rounds, have substantially boosted the anti-tank value of the Cockerill gun system.

133

MECHANISED COMBAT

Inside the Dragoon

The Dragoon APC is a straightforward 4×4 vehicle able to carry a section of infantry and armed with a pintle-mounted 7.62-mm machine-gun. The chassis serves as the basis for several different models: this is the cannon-armed recce version.

wheeled vehicle with good acceleration and light armour is far more appropriate for a quick reaction force defending an airbase than a tracked APC, which is slower and requires far more maintenance.

The Dragoon hull is of monocoque construction, fabricated from ballistic steel plate with welded seams. The hull, doors, hatches and even direct vision devices are protected against 7.62-mm×51 ball ammunition fired from right against the vehicle. The standard of protection equals or exceeds that of all known light armoured vehicles in the under 15-tonne (14.8-ton) class. Equipped as a basic personnel carrier, the Dragoon accommodates up to 10 personnel in the troop compartment, but it is also offered as a heavy weapons carrier, mortar carrier, communications vehicle and engineer/logistics vehicle.

Two- or four-wheel drive

Sitting in the driver's position at the front of the hull, you have the vehicle commander to your right and the instrument panel to your left. You use a conventional steering wheel and the Dragoon's automatic transmission employs five forward gears. Acceleration is good: 0-32 km/h (0-20 mph) in just under 10 seconds, and it can sustain a top speed of 102 km/h (63.4 mph) on a metalled road. A lever to the right of the steering wheel lets you select two- or four-wheel drive: the former is more fuel-economical for long journeys. The Dragoon has a delightfully piercing horn, but it will only operate when the light switch is activated for normal driving, so you cannot sound it accidentally during a supposedly silent watch.

The driver has three vision blocks and the commander's and driver's seats are power-adjusted, so you can drive with your head out of the hatch or sink out of view to drive closed down. Seat belts are provided for the commander and driver but not for the men in the troop compartment, who sit on individual seats hoping the driver can read the ground ahead of him. Like many fast-wheeled APCs, the Dragoon can travel across country faster than the men in the back can cope with. The spectacle of an armoured fighting vehicle tackling rough terrain at speed makes good advertising for the AFV, but an infantry section with headaches and vomit-soaked combats is of limited military value. At least the Dragoon's reasonably roomy troop compartment and sensible seating plan minimises the discomfort.

Electronic warfare models on exercise: the vehicle on the right carries a modified Arrowpointe 25-mm cannon turret fitted with a long-range day/night optical surveillance system.

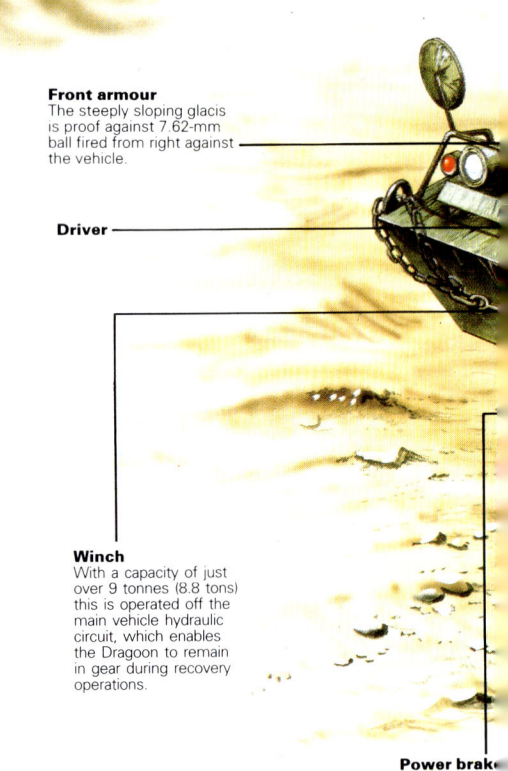

25-mm cannon
The M242 Bushmaster Chain Gun can be fired at 100, 200 or 500 rounds per minute. Recoil mechanism is internal and the spent cases are ejected forward of the vehicle.

Front armour
The steeply sloping glacis is proof against 7.62-mm ball fired from right against the vehicle.

Driver

Winch
With a capacity of just over 9 tonnes (8.8 tons) this is operated off the main vehicle hydraulic circuit, which enables the Dragoon to remain in gear during recovery operations.

Power brake

The Dragoon has a large door in each side which opens outwards, splitting in the middle so that the lower half becomes a step. The doors have rubber seals, as do the hatches, since the Dragoon is amphibious. A bilge pump system is fitted to cope with any leaks: three pumps, each capable of evacuating 190 l (50 gal) per minute. The amphibious capability is designed for rivers and lakes rather than the open sea. It should not enter the water if there are waves over 30 cm (11.8 in) high and, even in calm water, it paddles along at a sedate 4.8 km/h (3 mph) maximum. A propeller kit can be fitted which boosts water speed by about 50 per cent.

The engine compartment is at the rear of the Dragoon on the right-hand

MECHANISED COMBAT

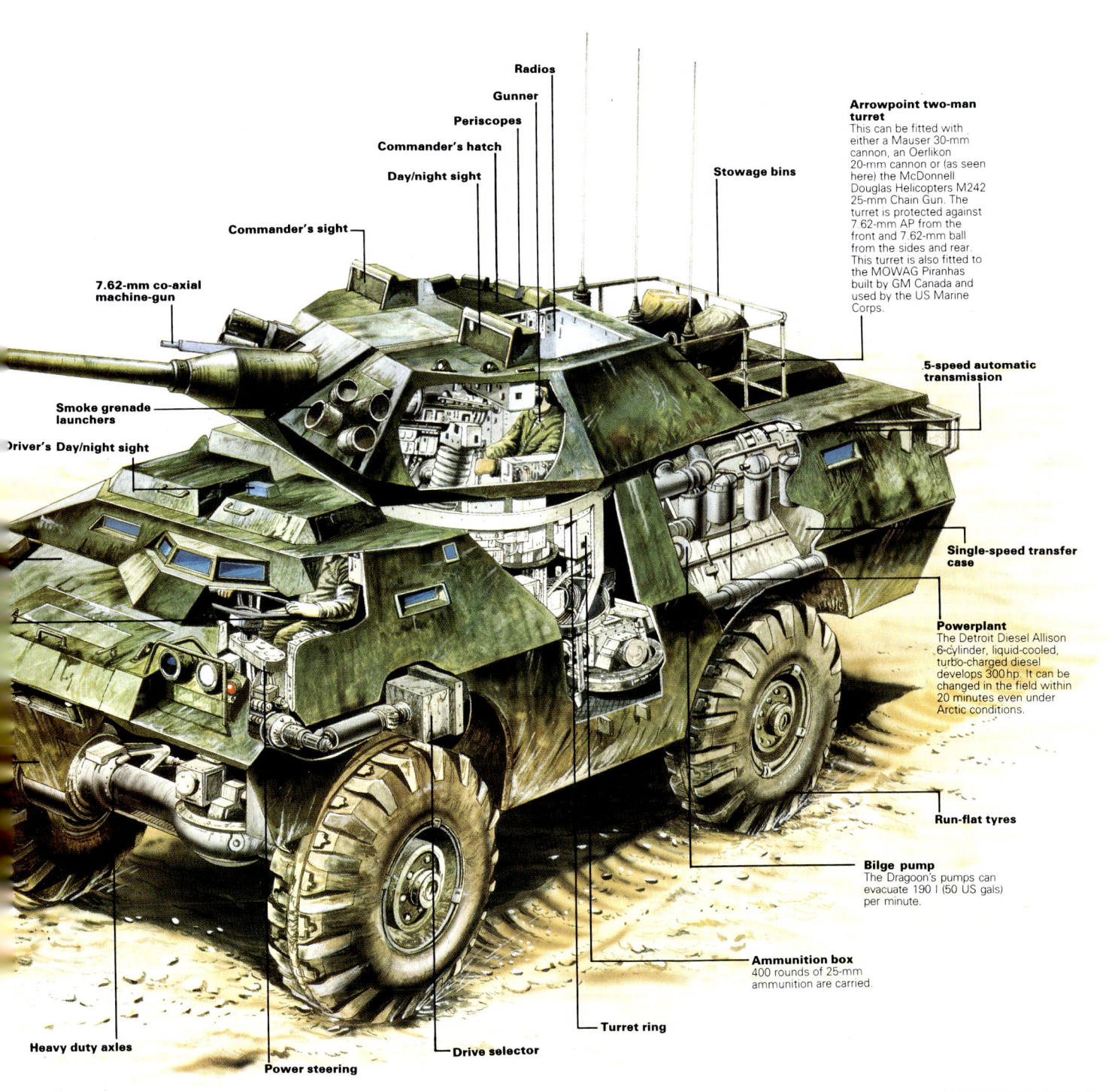

Arrowpoint two-man turret
This can be fitted with either a Mauser 30-mm cannon, an Oerlikon 20-mm cannon or (as seen here) the McDonnell Douglas Helicopters M242 25-mm Chain Gun. The turret is protected against 7.62-mm AP from the front and 7.62-mm ball from the sides and rear. This turret is also fitted to the MOWAG Piranhas built by GM Canada and used by the US Marine Corps.

Radios

Gunner

Periscopes

Commander's hatch

Day/night sight

Stowage bins

Commander's sight

7.62-mm co-axial machine-gun

Smoke grenade launchers

Driver's Day/night sight

5-speed automatic transmission

Single-speed transfer case

Powerplant
The Detroit Diesel Allison 6-cylinder, liquid-cooled, turbo-charged diesel develops 300 hp. It can be changed in the field within 20 minutes even under Arctic conditions.

Run-flat tyres

Bilge pump
The Dragoon's pumps can evacuate 190 l (50 US gals) per minute.

Ammunition box
400 rounds of 25-mm ammunition are carried.

Turret ring

Heavy duty axles

Drive selector

Power steering

side, and the exclusive 'powerpack system' is designed so that the engine transmission, cooling and associated systems can be removed in a single unit in under 20 minutes. This was demonstrated in the late 1980s by US Army National Guardsmen at Camp Grayling in northern Michigan's arctic temperatures.

Weapon load

The Dragoon APC is equipped with a 360-degree rotating machine-gun hatch carrying a 7.62-mm or .50-cal

Excellent reliability, fine cross-country performance and 60 rounds of 90-mm ammunition make the Dragoon Assault Gun a useful recce vehicle. The sharp slope of the glacis is clearly visible.

MECHANISED COMBAT

The Electronic Warfare Dragoon intercepts enemy raido signals, providing direction-finding and jamming facilities. The invisible war of the air waves may seem remote from the ground battle with rifles, but good EW capability can ensure that enemy command posts are stonked with artillery and their chain of command severely disrupted.

machine-gun protected by a small ballistic shield. Ammunition stowage is provided for 2,000 rounds, and five of the 10 vision blocks have weapon ports able to accommodate most hand-held weapons, including the M60 7.62-mm GPMG. The air-conditioning takes care of the gases produced on firing, but prolonged shooting from inside is not recommended.

The Dragoon assault gun variant offers the same performance characteristics as the standard APC, but has a two-man turret fitted with a 90-mm light gun and co-axial 7.62-mm machine-gun. The turret is fabricated from the same high hardness steel plate as the hull. Fire control is hydraulically driven and operated in azimuth and elevation by a single hand grip controller; back-up manual controls are provided in case of failure. The gunner uses the standard US Army M36E1 passive day/night sight, but thermal imaging and other sight options are available. M27 periscopes are provided for the commander.

Ammunition stowage

The assault gun's turret bustle can fit a variety of radio communications systems and stowage is provided in the turret for 10 90-mm shells and 500 7.62-mm rounds. Another 50 90-mm shells and 2,000 rounds of 7.62 can be fitted into the hull. The ammunition stowage is thus very generous but arguably dangerous: Israeli experience has shown that, since most hits on a tank are on the turret, all ammunition should be stowed below the turret ring. However, the Dragoon's armour will be penetrated by large-calibre cannon, so it is a largely academic question.

Mortar carrier

The Dragoon mortar carrier is an economic way to provide infantry with heavy weapons and high mobility. Retaining the same performance as the APC, it mounts an 81-mm mortar on a 360-degree rotating turntable that can be locked in a number of azimuth positions. Further fine tuning in azimuth and elevation is then possible to lay the mortar on target. Sixty mortar rounds can be stored within the vehicle and there is also room for aiming stakes, base plate and bipod, so the mortar can be dismounted and fired from outside the vehicle if required.

The Dragoon Maintenance and Engineer variant provides field recovery and support for the other Dragoons. Its 2,268 kg (5,000 lb) hydraulic boom crane serves to install or remove vehicle powerpacks or turrets, change tyres and load or unload cargo. Auxiliary power is provided for operating air and hydraulic accessories and tools.

The Dragoon series is completed by the Command, Control and Communications (C^3) variant, which can be finished to the customer's requirements and will accommodate numerous tactical communications and electronic warfare systems. From the outside, it looks little different from the basic APC, so unless you festoon it with tall aerials the enemy will not detect its true function. Inside it has a three-kilowatt auxiliary power generator, racks and mounts for communications systems and accessory equipment plus a camouflage net cover 166.8 sq km (1,796 sq ft), which can hide your command centre from prying eyes.

Field support

Given the plethora of wheeled APC families currently in service or offered for sale, the introduction of a new series may seem a bold or even foolhardy venture but the Dragoon is succeeding. It makes no extravagant performance claims and offers nothing in the way of superior speed, firepower or protection over its numerous commercial rivals. What it does offer is meticulous attention to field support and logistics. The removable powerpack system and the overwhelming use of readily-available commercial components make it a very attractive tactical support vehicle.

Right: A handful of Dragoons were acquired by the US Army in 1982 for use in the electronic warfare by the 9th Infantry Division High Technology Test Bed (HTTB).

The Dragoon Command, Control and Communications Countermeasures version with its mast extended. With the mast lowered the vehicle looks very similar to the ordinary APC so it is less likely to be singled out.

MECHANISED COMBAT

ZSU-23-4 THE DEADLY ZOO

Below: A ZSU-23-4 fires a 40-round burst, guided by its B-76 'Gun Dish' radar. This system was a key element in the air defence of Soviet tank and motor rifle regiments.

Historically, Soviet air defence units were always heavily influenced by the principles of firepower, surprise and mobility. Futuristic surface-to-air missiles are now integrated with more traditional anti-aircraft guns to develop a comprehensive and formidable all-round air defence coverage.

Shortly after the end of World War II, the Soviets introduced the ZPU-1 (single-barrel), ZPU-2 (twin-barrel) and ZPU-4 (four-barrel) range of light anti-aircraft guns, all of which incorporated the formidable 14.5-mm Vladimirov KPV heavy machine-gun, which can still be found mounted on several former Warsaw Pact and Third World armoured fighting vehicles.

Self-propelled AA

It was not until 1957, with the coming of the ZSU-57-2, that the Soviets could boast a truly modern purpose-built anti-aircraft system. Mounted on a shortened T-54 tank chassis and protected by a lightly armoured turret, the twin 57-mm guns of the ZSU-57-2 were derived from the blueprints of the German 5.7-cm 'Flakgerät' captured in the final stages of World War II.

The new Soviet weapon was capable of firing up to 120 rounds per barrel per minute to an effective altitude of 4,875 m (15,994 ft), but although the Soviets now possessed a weapon system capable of challenging any NATO fighter ground attack aircraft of the day, it was clear that the large 57-mm high-explosive ammunition was wastefully large.

More fundamentally, the system was manually controlled and therefore highly inaccurate. The necessity for a smaller-calibre, radar-controlled but highly-mobile replacement was placed high on the list of Soviet priorities.

The ZU-23 gun

A substantial breakthrough was made when the ZPU series of light anti-aircraft machine-guns was replaced by the ZU-23 twin 23-mm system with a greater range, higher velocity and much increased rate of fire.

In the late 1980s the ZU-23 saw extensive action in the Lebanon, where the Israelis captured a number from the PLO, mounted on the rear of elderly BTR-152 APCs and civilian trucks. The ZU-23 was also used against United States' forces in Grenada and Vietnam and against South African Forces (ground as well as air) in Angola. Most recently it was ex-

In the Soviet Army the ZSU-23-4s operated in pairs 500 m (1,640 ft) behind the tanks and APCs during an attack. In defence they were sited in a cluster wherever the terrain most favoured deployment.

137

MECHANISED COMBAT

Inside the ZSU-23-4

Introduced into the Russian army more than 35 years ago, the ZSU-23-4 outclassed contemporary NATO anti-aircraft guns, and improvements to its radar and fire-control computer have ensured that it is still an effective system in the late 1990s.

The ZSU-23-4's radar cannot track a target flying at under 60 m (196.8 ft) and the crew must aim the cannon using optical sights. Visual aiming is quicker than relying on radar guidance but much less accurate.

'Gun Dish' radar
This is the NATO codename for the B-76 radar fitted to the ZSU-23-4, which allows it to operate in all weathers and at night. It is a good tracking radar, difficult to detect or evade, although ECM pods provided to Israel by the USA were able to jam the radar of Egyptian ZSU-23-4s in 1973.

Turret armour
The turret is protected by a meagre 9 mm (0.35 in) of armour, which can be penetrated by 0.50 calibre machine-guns.

Ammunition
The ZSU-23-4's 23-mm ammunition comes in belts of 500 rounds with one Armour Piercing to every three High Explosive rounds. Both have a tracer base, and the AP round can penetrate 25 mm of armour at 500 m (1,640 ft) and 19 mm at 1,000 m (3,280 ft). Each ZSU-23-4 carries 2,000 rounds in 40 boxes of 500, and supply trucks follow about 1 km (0.6 miles) behind with another 3,000 rounds for each vehicle.

Driver
The driver has a separate compartment and he can raise a windscreen and wiper in front of himself when his hatch is open.

Splash plate
This prevents water surging up into the driver's compartment when the ZSU-23-4 is fording a stream. The vehicle can cope with water up to a metre deep.

tensively used by the Iraqi air defence system in the Gulf War.

Each ZU-23 has a five-man crew consisting of a commander, two gun layers and two ammunition loaders. The Number One Layer designates the target and his Number Two controls the firing foot pedal, simultaneously trying to keep the target in the crosshairs. The barrels, which are air-cooled, frequently overheat but can be changed in less than 20 seconds by the simple operation of a quick-release handle mounted above the weapon.

New version needed

Although the ZU-23 was clearly a great step forward it still had crucial shortcomings. It lacked the versatility of the self-propelled gun, relying totally on independent transport for mobility. Above all, it lacked radar.

The ZSU-23-4, or 'Shilka', first entered service in 1965. Equipped with four tried-and-tested ZU 23-mm guns, water-cooled for greater reliability, mounted on a converted PT-76 light tank chassis and linked to the B-76 ('Gun Dish') radar, the ZSU-23-4 combined firepower, accuracy and manoeuvrability.

Foreign users

Within a short period the ZSU-23-4 was in front-line service throughout the Warsaw Pact and had been exported to the major nations of the pro-Soviet Arab world. It saw extensive service in the Middle East campaign

Soviet ZSU-23-4s parade past Lenin's inscrutable gaze: nine separate versions of the vehicle have been identified, many from parade photos like this. Differences in stowage arrangements and cooling vents are the most obvious changes.

MECHANISED COMBAT

Quadruple AZP-23 23-mm cannon
Separated from the crew compartment by the armoured bulkhead, the four cannon have a cyclic rate of fire of 800-1,000 rounds per minute, per barrel. It can engage targets using one or two guns rather than all four. Targets are usually engaged with 40-round bursts.

Danger: runaway gun
The ZSU-23-4's cannon sometimes continue to fire while the turret traverses after the gunner has finished a burst. This is a disagreeable experience for ground troops nearby and is one of the reasons why ZSU-23-4s travel some distance away from troops they are supporting.

Mantlet
Protected by just 10 mm (0.4 in) of armour, this is vulnerable to heavy machine-gun fire, so although the ZSU-23-4 is a fearsome weapon against ground troops its lack of protection makes it very vulnerable and the Russians are unlikely to risk it in this way against an enemy with plentiful heavy weapons.

Hull armour protection
The ZSU-23-4 has only 15 mm (0.6 in) of armour on its hull front, sloped at 55 degrees, and 15 mm (0.6 in) of armour on the hull sides. This protects the hull from small-arms fire and shell splinters but is easily penetrated by anti-tank rockets and cannon.

Chassis
Similar to that of the PT-76 light tank, the chassis of the ZSU-23-4 has an overpressure NBC system but, surprisingly, is not amphibious. It crosses rivers on GSP ferries and can fire while afloat.

of October 1973, when it was credited with the destruction of 31 of the 110 Israeli aircraft downed. During the past 20 years it has seen considerable service with considerably less success, although this probably has more to do with the Iraqis' inability to maintain the radar than to the diminution of the equipment.

The hull is of all-welded construction with the driver at the front, three-man turret in the centre, and engine and transmission at the rear. The model V-6 six-cylinder inline water-cooled diesel engine, which develops 240 hp at 1,800 rpm, is in effect one half of that fitted to the old T-54 Main Battle Tank, and incorporates a pre-heater to make starting easier in the freezing Russian winters.

Turret

The turret, with power traverse and elevation, has a maximum elevation of +80°, depression of −7° and total traverse of 360°. It is spacious, allowing the commander, gunner and radar operator to work in comparative comfort, and is fully NBC (Nuclear, Biological and Chemical) sealed.

The gas-operated 23-mm cannon, with its vertically-moving breech-

Soviet AFV crews were not cross-trained: this driver could not use the vehicle's weapons, operate the radio or read a map. On the other hand, he can cope very well with the 'soldier proof' engine.

MECHANISED COMBAT

block locking system, is similar to that mounted on the original ZU-23. Each of the four guns has a cyclic rate of fire of 800 to 1,000 rounds per minute, the greatest of any anti-aircraft gun in present day service, but the ZU-23-4 normally fires in bursts of 50 rounds per barrel, reducing its effective rate of fire to 200 rounds per barrel per minute.

Rounds and range

Both HEI (High-Explosive Incendiary) and API (Armour-Piercing Incendiary) rounds, weighing 0.19 kg (6.7 oz) and 0.189 kg (6.6 oz), respectively, are carried and can be fired to a maximum horizontal range of 7,000 m (22,965 ft) and vertical range of 5,100 m (16,732 ft). Effective anti-aircraft range is, however, between 2,000 and 2,500 m (6,562 and 8,200 ft), and effective range in the ground role is 2,000 m (6,560 ft).

The B-76 radar, known throughout NATO by the code name 'Gun Dish', has both target acquisition and fire control capabilities. The radar scanner, which takes the form of a parabolic dish, is positioned to the rear of the turret but can be rotated and folded down when not in use. However, it must remain upright and clear when the vehicle is operational, making camouflage virtually impossible.

Operating on the Ku band, the radar can pick up aircraft up to 20 km (12.4 miles) away, after which the MTI (Moving Target Indicator) acts as an analogue computer locking on to the target. Power is supplied to the radar by an auxiliary gas turbine for added reliability, but optical sights are also provided, although these are so inaccurate as to be virtually useless.

Although 'Gun Dish' suffers from 'clutter' when tracking targets below 60 m (196.8 ft), it is nevertheless regarded as so dangerous that NATO pilots will try to avoid flying within range and still regard it as one of the most, if not the most, lethal anti-aircraft weapon system in existence.

Elevated to 80°, it is easy to see from this photograph why the ZSU-23-4 was used in Afghanistan to engage guerrillas firing down from the mountain tops. However, in frontier arithmetic the cost and weak armour of the vehicle counted against it.

ZSU-23-4s paraded by the Libyan leader Colonel Gaddafi. With a foreign policy likely to lead to further visits by the US Air Force, Libya's need for anti-aircraft weapons has never been greater.

NATO

The United States' attempt to produce a cheap, off-the-peg, self-propelled system met with total disaster. An attempt was made by Ford to link the existing 40-mm Bofors gun with the old M48 tank chassis, and even more optimistically with the F-16 fighter aircraft radar. The system, known as Sergeant York, proved totally impractical and was eventually abandoned by the Pentagon with severe loss of face and finance.

Despite its age, the ZSU-23-4 still constitutes a very potent anti-aircraft weapon. It was issued to the Soviet army in the scale of four per regiment, where it operated in pairs in close support headquarters protection. Its simple valve technology and robustness make it a favourite of the gunners who work with it, and it is likely to remain in front-line service with the CIS and former Soviet client states for many years to come.

Interestingly, the Soviet Union trialled a new system in the years before the fall of the Soviet empire. The new vehicle is larger than the ZSU-23-4, mounts 30-mm guns and short-range surface-to-air missiles on a common turret, and bears a striking resemblance to the German Gepard.

MECHANISED COMBAT

SAXON THE SLAYER

Seen here with a turret fitted with twin 7.62-mm machine-guns, Saxon is a versatile wheeled APC which is playing an increasingly important role in the British Army.

Saxon's mobility is obviously not as good as that of an MICV like Warrior but many armies, including the Russian, have found wheeled APCs perfectly adequate.

Saxon's true forte is IS (Internal Security) work. The design takes account of the long experience of the British Army in Northern Ireland and has attracted interest worldwide. In the Middle East, Saxon has been purchased by Bahrain.

The Saxon AT105 is a rugged, no-nonsense, multi-purpose, wheeled armoured personnel carrier. It is in service with the British Army and a number of other armies and police forces throughout the world, for example, Bahrain, Nigeria, and Malaysia, where 140 are in service with the Police Field Force. The straightforward 'no frills' design means that it can be adapted for almost any use. The standard version does not have a revolving turret and has no fixed armament, but the range offers various degrees of sophistication. The Saxon is suitable as a conventional APC, as an internal security vehicle or as a police support vehicle.

Saxon is mobile and manoeuvrable, carrying up to 12 men at speeds of up to 96 km/h (59.6 mph). The automatic gear-box is combined with a transfer box that gives two- or four-wheel drive and high or low ratios in all gears. Experience has shown that the Saxon is a very impressive cross-country performer.

Saxons in the standard APC configuration can be supported by Saxons adapted as command-post vehicles, ambulances, communications vehicles of various kinds, and hardened OPs for artillery forward observation officers. Together, they form a 'family' of vehicles, of simple construction and using standard components. This

MECHANISED COMBAT

Inside the Saxon

Saxon is a very straightforward wheeled APC, economical to run and easy to maintain. It will equip many of the infantry battalions based in the UK but is earmarked for BAOR in time of war. The chassis can easily be modified to carry heavy weapons or to operate as a command post, ambulance or other specialist vehicle.

Sheet armour
The cupoloa is bolted to a sheet of armour so that the whole assembly can easily be removed and replaced by one with a different weapon fit.

7.62-mm GPMG

Commander's cupola

Padded bench sea
These can be fitted w seat belts.

Driver
Saxon is available in both right- and left-hand drive. Steering is power assisted and the vehicle has an automatic gearbox.

Bullet-proof windscreen

Wing mirrors

All-welded steel hull
This is proof against not just 7.72-mm ball but will keep out armour-piercing bullets as well. It will stop shell fragments from a 155-mm shell detonating 10 m (32.8 ft) away.

Protected headlights

The Saxon Recovery Vehicle carries a Hudson Wharton capstan 5-tonne (4.9-ton) hydraulic winch able to cope with vehicles weighing up to 16 tonnes (15.7 tons). In British service they are operated by a four-man REME team.

Multi-leaf semi-elliptical springs

Mine-resisting floor
Many wheeled APCs based on existing truck chassis are vulnerable to mine damage, but the V-shaped floor of Saxon provides maximum protection for the occupants.

means that parts can be obtained easily and relatively cheaply – ensuring a high level of operational readiness.

Riot control

But, more than anything else, the Saxon is an ideal internal security (IS) vehicle. It was designed to meet all requirements for anti-riot work in urban areas, as well as for counter-insurgency and guerrilla warfare. The hull will withstand 5.56-mm and 7.62-mm armour-piercing or ball rounds at point-blank range. Its special shape defeats mine blast pockets and gives

MECHANISED COMBAT

Exterior stowage bins
In addition to saving interior space, the external bins provide a little extra protection against small-calibre hollow charge anti-tank weapons. Excluding the roof, Saxon has 1.56 m³ (55 ft³) of stowage space.

The Saxon command vehicle can be fitted with a turret mounting one or two GPMGs and the interior is extensively modified. The right-hand side of the hull is fitted with staff positions and there is a map board on the left.

Run-flat cross-country tyres
Once shot through, the tyres will still carry Saxon for 96 km (59.6 miles) at up to 48 km/h (30 mph).

maximum protection to the crew, engine, gear-box and radiator. 'Run-flat' tyres are standard.

You will find that it's very easy to get in and out of the Saxon, which is particularly important in IS work. In an ambush, you'll find the two large doors at the rear, which are controlled by the driver, and the two side doors, very comforting. There are six observation or firing ports, and the driver has four small windows fitted with bullet-proof glass giving the same degree of protection as the armour plate. The vehicle can be adapted for a number of IS purposes, including that of explosive ordnance disposal (EOD) and crowd control. The latter model can be fitted with a barrier remover, below the radiator, to cope with any obstacles put in its path, and with folding 'wings' to protect troops alongside the vehicle from rioters' missiles.

Saxon, manufactured by GKN Sankey of Telford in Shropshire, was put through exhaustive trials during its development. As well as 7.62-mm and 5.56-mm rounds fired point blank at the hull, a range of weapons – including petrol bombs, nail bombs and incendiaries – were tested against it.

On one occasion, the fireball from a blast incendiary completely enveloped the test vehicle – with no major external damage, and no evidence of any internal effects. Claymore mines, surface mines and pressure mines were also tried out against the vehicle. While there was external damage in all the tests, there was no significant internal damage. All the occupants would have been fully protected.

Tough and rugged

During an anti-tank mine test, a wheel assembly was sheared and the entire vehicle was moved two metres sideways. Again, the crew would have been unharmed, if somewhat shaken. There is no doubt that the Saxon is a genuinely tough and rugged vehicle, in which you can be perfectly confident that you will be properly protected.

Optional interior arrangements

The British Army Saxon APCs differ from those sold abroad by having external stowage bins and no door in the left-hand side of the hull. Each door had a firing port and vision block and there are additional firing ports along the hull side.

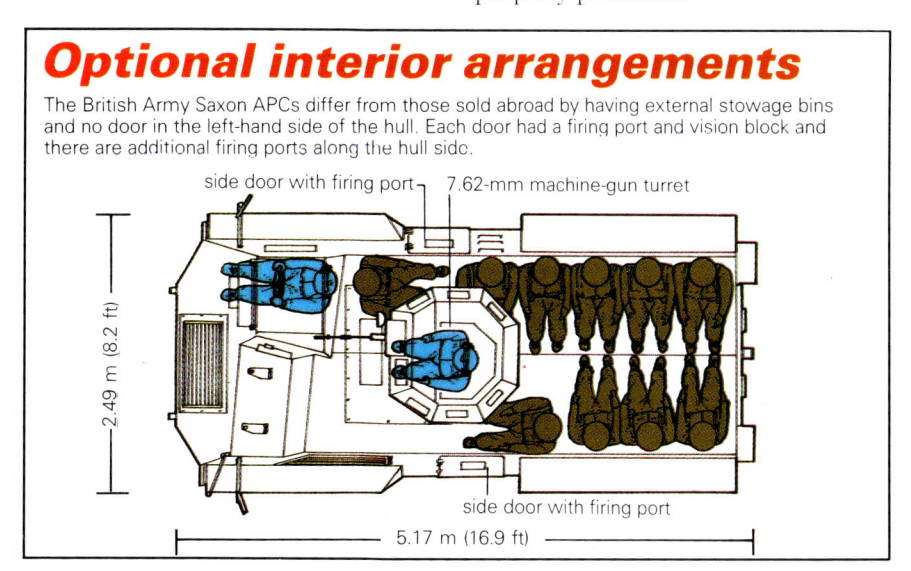

side door with firing port — 7.62-mm machine-gun turret

2.49 m (8.2 ft)

side door with firing port

5.17 m (16.9 ft)

MECHANISED COMBAT

A Saxon shows off its IS equipment: the 'wings' of steel mesh to protect dismounted personnel have been in service for some time on Humber 'Pigs' in Northern Ireland. Note the barricade remover and mesh over the windows.

The Saxon is also suitable for police work, and is used for that by some Persian Gulf states. In this configuration it is painted white and is fitted with searchlights, flashing blue lights, sirens, and so on. For a paramilitary police force such as the French Gendarmerie or the German Bundesgrenzschutz, the Saxon makes an ideal vehicle. There are not many APCs that can be used as a conventional military APC, an IS vehicle and a police immediate response vehicle.

Inside the Saxon

The section commander sits immediately behind the driver, and doubles as a gunner if the vehicle is attacked. Although a machine-gun is not standard, a pintle-mounted machine-gun can be fitted in the commander's cupola, or a revolving 7.62-mm GPMG turret can be mounted. In this event the commander has his hands full directing the driver and operating the machine-gun, so the best solution then is for the section second-in-command (who becomes the *vehicle* commander) to occupy the turret, leaving the section commander to command the section.

You can fight from the vehicle by firing through the ports on either side, but British Army tactics generally call for you to disembark from the vehicle short of the objective and assault on foot. That is why the commander should really be in the back of the vehicle so that he can lead the section into the attack.

A Saxon ambulance crosses a ditch filled in with plastic pipe fascines. The ambulance version can accommodate four stretcher cases and a couple of orderlies in addition to commander and driver.

The rear compartment can accommodate between eight and 10 fully-equipped men on padded benches – you'll find that eight is more comfortable. If you leave the section second-in-command in charge of the vehicle, you can be sure it will be in good hands and in the right place when you need it again. In the attack, when you have debussed, he will manoeuvre the vehicle into a hull-down position to give you covering fire during the assault.

If you are posted to one of the British BAOR reinforcement brigades, you'll be part of a Saxon crew. Even though it's an ugly beast, you'll soon be won over by this vehicle. It's easy to drive: vision is good, power steering is standard, you have an automatic gear-box and the engine – a standard Bedford six-cylinder diesel – won't let you down.

Even more important, Saxon will protect you. The armour works. Most important of all, it will get you there. With a range of 480 km (298 miles), run-flat tyres and a road speed of 96 km/h (59.6 mph), Saxon eats up miles. It's worth learning about this vehicle, as you could find yourself operating in one. Even in Northern Ireland, the 'Pigs' can't last for ever. What's the betting Saxon will take over?

MECHANISED COMBAT

ON THE LINE WITH THE
M109

In service for 40 years, the M109 155-mm self-propelled gun equips artillery regiments throughout NATO.

Well used artillery causes more casualties to the enemy and has a greater adverse effect on his will to fight than any other weapon. No matter how capable the latest tanks and infantry weapons may be, neither can hope to bring about victory if the enemy can bring down accurate artillery fire on their positions. But if artillery is to carry out its prime role of disrupting, demoralising and destroying the enemy, it must have sufficient range and power to bring down concentrated fire on enemy positions.

The artillery's job has been made more difficult by the introduction of much improved tank armour. Less than fifteen years ago a near miss from a 105-mm Abbot would have immobilised an enemy tank, but nothing less than a 155-mm gun can hope to have that effect against T-64s, T-72s or T-80s. Fortunately the venerable but reliable United States' M109 howitzer adequately fills this role.

Production
The first production model of the M109 was completed by the Cadillac Motor Car Division of General Motors in 1962. Since then some 3,700 models have been completed, 1,800 for United States Army service, the rest for export to over 15 countries throughout Europe and the Middle East, making the M109 the most widely used self-propelled howitzer in the world.

Continual improvements to the original design are most obviously seen in the longer M185 gun, first seen on the M109A1.

Design and equipment
Although the all-welded aluminium hull and turret provide the crew with protection against small arms fire, they are of little use against shrapnel. The comparatively short-ranged M109 would be vulnerable to counter-

145

MECHANISED COMBAT

battery fire in an artillery duel with its Soviet divisional level equivalent, the 2S3.

Firepower

The stubby 155-mm L/23 howitzer has a distinctive smoke extractor approximately two-thirds of the way along the barrel. The semi-automatic breech block and hydro-pneumatic recoil system enable an astounding rate of fire of three rounds per minute to be attained over a short period. However, the 43-kg (94.8-lb) shell is difficult to handle in the confined turret, and crew fatigue reduces the practical rate of sustained fire to 45 rounds per hour. Furthermore, as each gun carries only 28 rounds of ammunition, replenishment soon becomes a problem.

Variants

Germany, Italy and Switzerland have produced their own variants of the original M109, in each case together with an improved gunnery system. The West German M109G has a Rheinmetall horizontal sliding breech block and locally-produced sights, enabling it to fire domestically-produced ammunition to a much increased range of 18,500 m (60,695 ft). The Italian variant, built under licence by OTO-Melara, has a lengthened barrel to accept ammunition originally designed for the FH-70. Using standard ammunition this has a range of 24,000 m (78,740 ft). The Swiss, with an eye to fire and movement, rather than range, have fitted a semi-automatic loader to their variant, designated M109U, to increase the rate of fire to six rounds per minute.

Attempts were made throughout the 1970s to build a joint European replacement for the M109. The new gun, provisionally named SP-70, with its Leopard 1 chassis and FH-70 barrel was to have been able to outrange and outperform all but the heaviest of Soviet guns. Unfortunately national self-interests prevailed and production was regularly postponed until the whole concept was finally abandoned late in the 1980s.

M109A1/A2

From the outset it was obvious that measures would have to be taken to update the M109 to provide a stop-gap until the introduction of the SP-70. To this end the M109A1 was introduced, the first model entering service with

Inside the M109

The M109 played a key role in the defence of western Europe and latterly in the Gulf War. Highly mobile to avoid enemy counter-battery fire, its 155-mm gun provides vital fire support for NATO infantry. Used by most NATO armies, the example illustrated here is in US Army markings. In the background another M109 replenishes its ammunition supply from an M992 Field Artillery Ammunition Support Vehicle (FAASV).

A US Army M109 with all hatches open to give some ventilation in the still air of the desert. US forces in the Gulf War were embarrassed to find Iraq's Soviet-made artillery had longer range than most NATO guns.

MECHANISED COMBAT

M185 155-mm howitzer
This can elevate to + 75 degrees and has a maximum range of 18 to 24 km (11.2 to 14.9 miles) depending on the type of ammunition fired. It fires various HE rounds, smoke, illumination, mines and bomblets, CLGP and binary chemical shells. This longer-barrelled version was introduced on the M109A1 and increased the range by 3,300 m (10,826 ft).

.50-cal Browning machine-gun
This is pintle-mounted on the front of the commander's cupola on the right-hand side of the turret. Its primary function is for anti-aircraft fire; if the commander finds himself engaging enemy infantry with it, the M109 is in serious trouble.

Rear access door
This has to remain open when in action because the turret quickly fills with fumes despite the new fume extractors fitted on the M109A1. The M109A5 has an NBC system to properly ventilate the turret, so it can operate 'closed down' without suffocating the crew.

the British Army in 1978.

The M109A1 is, in essence, the M109 fitted with the much longer M185 155-mm cannon. This new gun has a very effective fume extractor preventing propellant gases from entering the turret after firing, takes a bigger charge and offers a greater maximum range of 18,000 m (59,055 ft) compared with 14,700 m (48,228 ft) of the standard M109.

The M109A2 has an improved shell rammer and recoil mechanism, an M178 modified gun mount and other minor improvements. To date there are 179 M109A1s/A2s in service with

The stubby barrel of the original M109. Twenty-three calibres long, it has a substantial fume extractor behind the muzzle brake. At the time of its introduction in NATO, the Russians still relied on towed artillery.

147

MECHANISED COMBAT

Royal Artillery Medium Regiments.

The future of the M109 was assured in September 1985, when the United States Government awarded a $53 million contract for the production of the M109A5 under its Howitzer Improvement Program (HIP). Under that contract, 11 existing M109s are being converted, nine for the US Army and two for Israel.

Numerous improvements are being made to the M109 HIP. A new aluminium armoured turret will have additional storage for 36 charges, while a combination of a new Emerson Electric automatic loader and Honeywell modular automatic fire-control system allow for burst fire of three rounds in 15 seconds, followed by a sustained rate of eight rounds per minute – far superior to anything in present-day service. Equally important for crew comfort, the NBC protection system will operate during firing; at present the M109 is forced to fire with its rear door open. The M109HIP will mount the improved 155-mm, 39-cal gun, although this may be replaced in the future by a 45-cal tube with an estimated range of 38,000 m (124,670 ft), greater than the majority of 203-mm heavy guns in existence. This, however, remains for the future, as the much larger and more powerful gun will require a reinforced chassis and suspension.

The M109 is far less costly than any of its counterparts. It has enabled standardisation of medium artillery within NATO, and has offered many countries which would not otherwise have been able to afford it the possibility of purchasing a large arsenal of self-propelled guns, and to modernise them at minimum cost.

Above: A US Army M109 on winter exercises in Germany. NATO M109s are trained to 'shoot and scoot': firing for short periods before moving off to avoid enemy counter-battery fire.

The US Army uses the specialist M992 FAASV to support its M109s. The M992 can feed 155-mm shells into an M109 along a special conveyor belt at a rate of about eight rounds a minute.

MECHANISED COMBAT

SCOUTING
with the Saladin

Iraqi tanks were passing along the far end of the street. A six-wheeled armoured car roared around the corner and lurched to a halt. Its stubby gun elevated slightly and fired, the deafening report echoed around the high-rise buildings from which frightened observers witnessed this unequal duel. But before the tanks could take their revenge, the armoured car spat out a volley of rockets that wrapped it in a protective blanket of smoke. Concealed by the billowing white cloud, it reversed rapidly and trundled down a side alley.

This armoured car's solo counter-attack was filmed as the Iraqi forces poured into Kuwait City in August 1990. The vehicle was not new, unlike the gleaming tanks of the Republican Guard. Designated FV601 in the British Army, the Saladin armoured car was widely exported to the Middle East and Africa. With its six large wheels and chunky silhouette, it may look like an outsize Tonka toy, but the Saladin's 76-mm gun was a giant-

A British Army Saladin in pristine condition with the commander's hatch open and the 76-mm gun elevated. The yellow circle below the smoke discharger bank was used to detect chemical weapons.

killer in its day and is still a serious menace to all but the new generation of Main Battle Tanks.

The Saladin was designed after World War II to replace the British Army's range of armoured cars armed with machine-guns and diminutive 2-pdr guns. The British had spent six years fighting the German army with

MECHANISED COMBAT

Above: Kuwaiti Saladins are shown in action against Iraqi armour during the first hours of Saddam Hussein's invasion. Note the empty cases from the 76-mm shells lying on the pavement.

Below: Crew members inspect a British Army Saladin that was badly damaged by a terrorist attack in Aden. Saladins were also used on counter-insurgency operations in Cyprus during the war against EOKA.

its excellent series of six- and eight-wheeled armoured cars. Now they followed suit. Saladin's chassis was very similar to that of the Army's new armoured personnel carrier, the FV603 Saracen. Indeed, the armoured car was delayed while Saracens were rushed off Alvis' Coventry production line to join the British forces fighting the communist guerrillas in Malaya.

Production finally began in 1958 and continued until 1972 as export orders continued to come in. Operated by a three-man crew, the Saladin proved to be a tough and reliable vehicle with an impressive cross-country performance. The vehicle is divided into three compartments: the driver's at the front, the engine at the rear and the fighting compartment in the middle. The driver sits in a central position with a single-piece hatch cover that folds forward onto the hull front. In action this has to be closed, the driver relying on three periscopes, one in the hatch cover and one either side.

The Saladin weighs 11.5 tonnes (11.3 tons) and its Rolls-Royce eight-cylinder petrol engine can reach 72 km/h (45 mph). As befits a reconnaissance vehicle that may need to get out of trouble in a hurry, all five gears can be used in either direction – although a full-speed reverse can be a fraught experience. It can tackle gradients of up to 46 per cent and ford up to a depth of 1.07 m (3.5 ft). The engine is separated from the fighting compartment by a fireproof bulkhead. The driver has a warning device to alert him to a fire in the engine compartment, and he can operate a fire extinguisher system from his position.

The commander's position is on the right-hand side of the turret, with the gunner on the left. The commander has four periscopes facing forwards and a swivelling periscope to cover the rear arc. He has to double as a loader, feeding 76-mm shells into the L5 gun. A total of 42 rounds is carried for the main armament: 11 between the commander and the gunner, 11 on the right behind the driver, 12 on the left, and eight in the hull rear.

The gunner has two periscopes; the lower one has ×6 magnification and is used to aim the gun. When it was designed, a vehicle the size of Saladin could not mount the same sort of gun as a tank. Instead, Saladin carried a short-barrelled gun optimised for HESH (High-Explosive Squash Head) projectiles rather than high-velocity armour-piercing rounds.

HESH round

The 76-mm HESH round weighs 7.4 kg (16.3 lb), including its brass cartridge case. When it strikes an enemy vehicle the soft explosive filling 'pancakes' onto the surface and detonates. Unless the vehicle is only lightly armoured – like another armoured car – the HESH round will not actually blow a hole in the target. But it still inflicts enormous damage. When over 5 kg (11 lb) of high explosive detonates against the outer surface of a sheet of armour, it causes a tremendous shock wave to pass through the metal. This blows huge scabs of armour off the inside, showering the target's interior with razor-sharp fragments.

The HESH round is capable of caus-

MECHANISED COMBAT

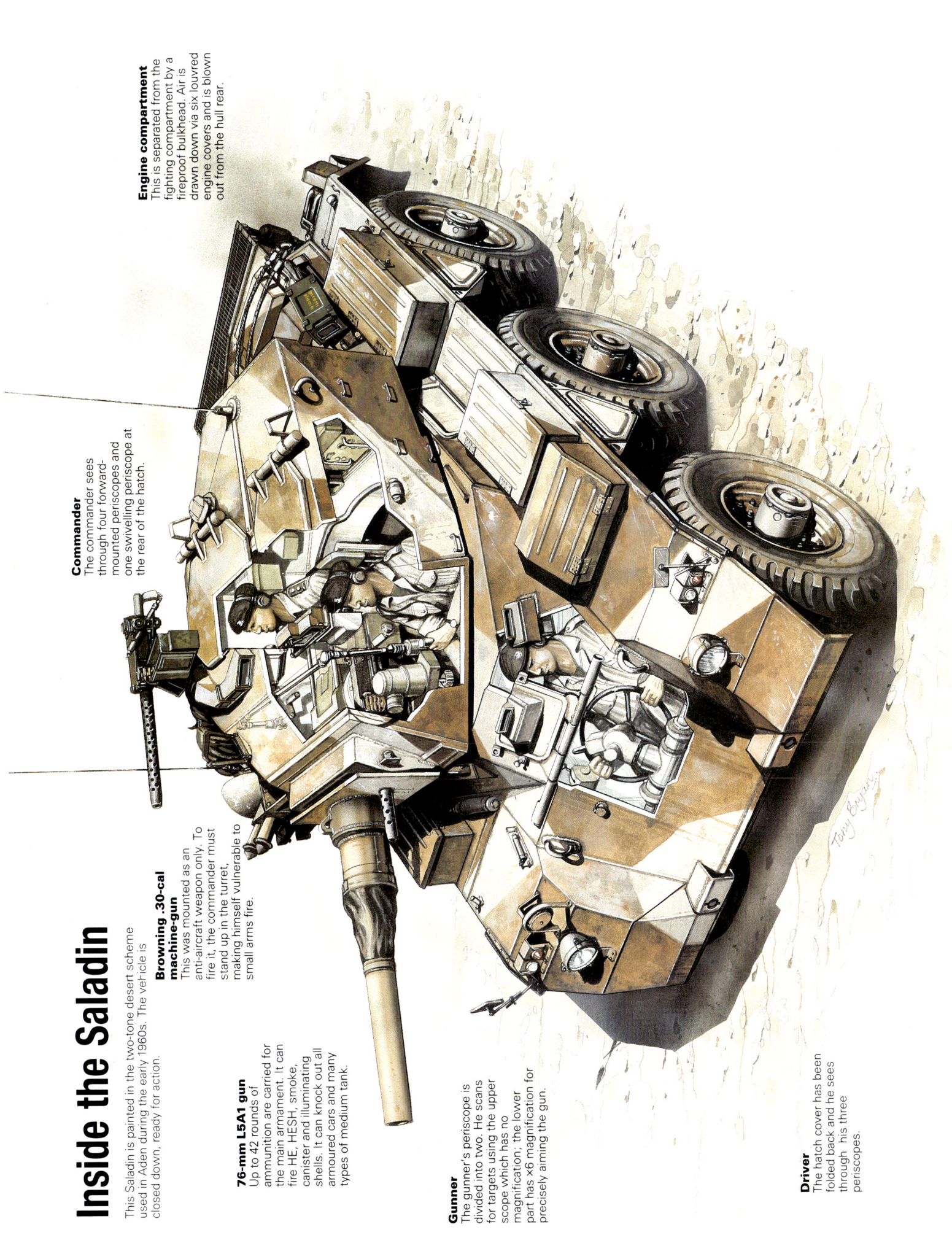

Inside the Saladin

This Saladin is painted in the two-tone desert scheme used in Aden during the early 1960s. The vehicle is closed down, ready for action.

76-mm L5A1 gun
Up to 42 rounds of ammunition are carried for the main armament. It can fire HE, HESH, smoke, canister and illuminating shells. It can knock out all armoured cars and many types of medium tank.

Gunner
The gunner's periscope is divided into two. He scans for targets using the upper scope which has no magnification; the lower part has ×6 magnification for precisely aiming the gun.

Browning .30-cal machine-gun
This was mounted as an anti-aircraft weapon only. To fire it, the commander must stand up in the turret, making himself vulnerable to small arms fire.

Engine compartment
This is separated from the fighting compartment by a fireproof bulkhead. Air is drawn down via six louvred engine covers and is blown out from the hull rear.

Commander
The commander sees through four forward-mounted periscopes and one swivelling periscope at the rear of the hatch.

Driver
The hatch cover has been folded back and he sees through his three periscopes.

151

MECHANISED COMBAT

ing serious damage to Russian tanks like the T-54/55, T-62 or the first model T-72. It is extremely dangerous to lightly protected MBTs like the AMX-30 or Leopard 1. This made the Saladin particularly attractive to armies in Africa and the Middle East that were likely to face older tanks.

Gun ammo

The 76-mm gun also fires HE, illuminating, practice and smoke rounds. A canister round was also manufactured but is no longer in production. This converted the gun to a massive shotgun; hitting enemy infantry with an unselective blast of fragments was a useful antidote to a jungle ambush.

British forces employed the Saladin in numerous guerrilla wars from the Far East to Cyprus and the Middle East. The angular hull of the vehicle anticipated later mine-protected vehicles, helping to divert the blast of a land-mine away from the vehicle. It was not unknown for a Saladin to drive back to base with only five wheels.

German police variant

There was only one variant of the Saladin: the FV601(D) which was used by West Germany's paramilitary border police, the *Bundesgrenzschutz*. This had no co-axial machine-gun and had different smoke rocket launchers and other minor changes. Some of these vehicles were sold to the Sudan during the 1960s, but there has been little sign of them in the current civil war there. The Australian army also used the Saladin, and it created an unusual hybrid when it withdrew them from service. The turrets were removed and fitted to American M113 armoured personnel carriers to create a 'Fire Support Vehicle'.

With no night vision equipment or NBC protection, the Saladin is at a serious disadvantage compared with modern armoured cars. But as a basic reconnaissance vehicle it is still effective. Its cross-country performance is quite creditable and it is not costly to run. Production ceased in 1972 after 1,177 were built, and they will no doubt continue to serve for some years to come.

Inset: Indonesia is one of many countries that still operates the Saladin. Cheap to run and uncomplicated to maintain, the Saladin was widely exported to Africa and Asia.

Main picture: One of the last Saladins in service with the British Army surrounded by coloured smoke. The turret and gun are shrouded in camouflage netting.

MECHANISED COMBAT

BATTLING WITH THE BMP

The BMP was built to allow Soviet infantry to keep attacking at high speed even during a nuclear war. Combined forces of Main Battle Tanks and infantry in BMPs were trained to smash through the enemy defences and continue to advance at a cracking pace 70 to 100 km (43.75 to 62.5 miles) a day. The fast-moving Russian armour is never halted long enough to be targeted by tactical nuclear weapons.

When the BMP (Boyevaya Mashina Pekhoty) made its first public appearance in Moscow at the October Revolution Parade of 1967, it immediately became clear that the Warsaw Pact now possessed an armoured personnel carrier superior in every respect to anything in the NATO arsenals.

Although the BMP was small by Western standards, its 280-hp six-cylinder engine was nevertheless powerful enough to match the latest Soviet Main Battle Tanks in cross-country performance, and its crew compartment large enough to accommodate an eight-man section of fully-equipped infantrymen.

The Red Army learned from the bloodbaths of 1943 and 1944 that massed tank attacks invariably end in failure unless supported by infantry capable of exploiting any breaches in the enemy defences. At that time, rifle sections were carried forward into battle on the unprotected hulls of T-34 tanks with inevitable heavy losses.

After the war, Soviet engineers designed a series of wheeled APCs cap-

The majority of Russian front-line infantrymen ride into battle in the BMP mechanised infantry combat vehicle. Light and fast, the BMP is fully amphibious although its low silhouette makes it easily swamped in choppy water.

MECHANISED COMBAT

Inside the BMP

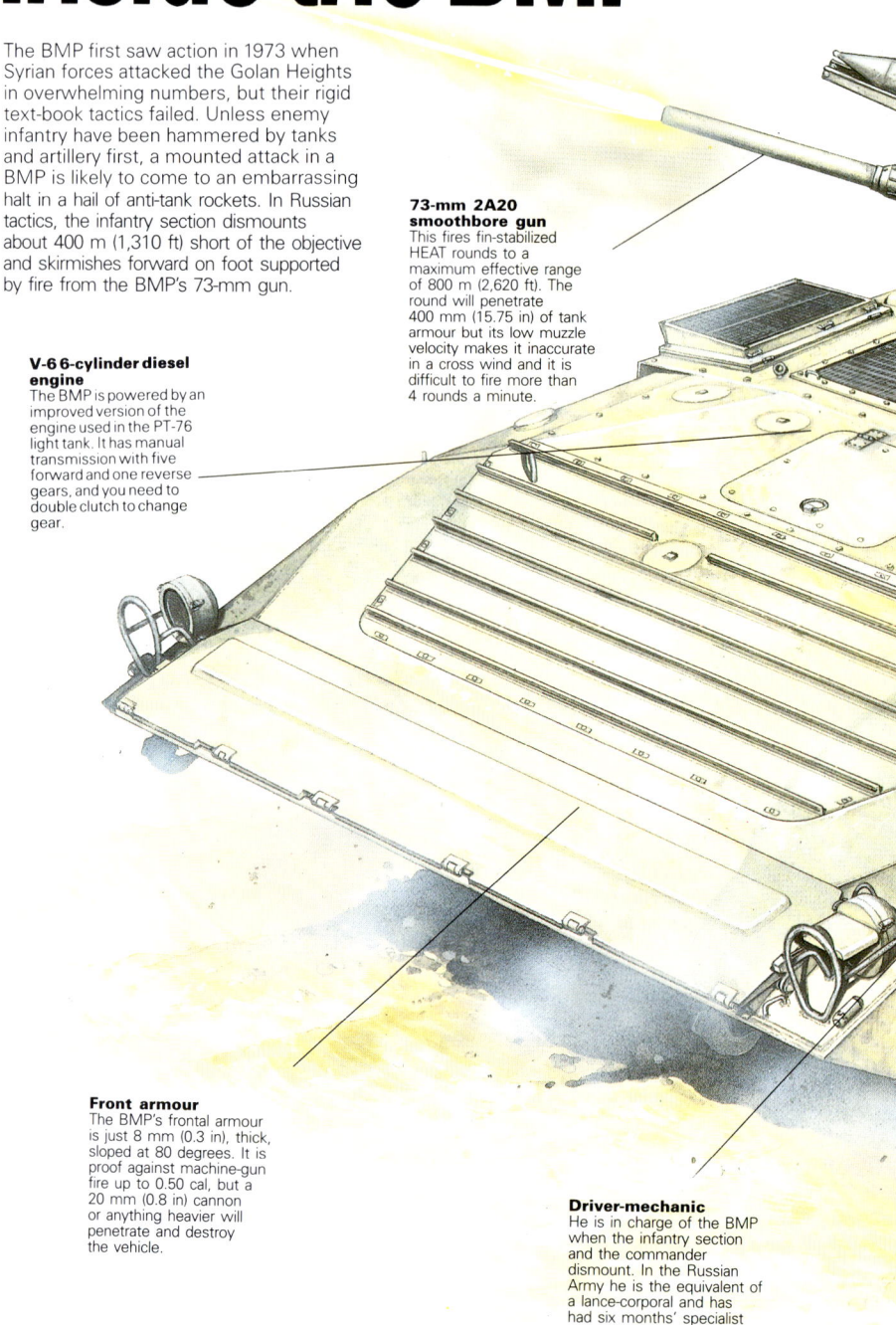

The BMP first saw action in 1973 when Syrian forces attacked the Golan Heights in overwhelming numbers, but their rigid text-book tactics failed. Unless enemy infantry have been hammered by tanks and artillery first, a mounted attack in a BMP is likely to come to an embarrassing halt in a hail of anti-tank rockets. In Russian tactics, the infantry section dismounts about 400 m (1,310 ft) short of the objective and skirmishes forward on foot supported by fire from the BMP's 73-mm gun.

The BMP's low silhouette makes it a smaller target, but it can only depress its gun by 4°, limiting its ability to use hull-down positions.

73-mm 2A20 smoothbore gun
This fires fin-stabilized HEAT rounds to a maximum effective range of 800 m (2,620 ft). The round will penetrate 400 mm (15.75 in) of tank armour but its low muzzle velocity makes it inaccurate in a cross wind and it is difficult to fire more than 4 rounds a minute.

V-6 6-cylinder diesel engine
The BMP is powered by an improved version of the engine used in the PT-76 light tank. It has manual transmission with five forward and one reverse gears, and you need to double clutch to change gear.

Front armour
The BMP's frontal armour is just 8 mm (0.3 in), thick, sloped at 80 degrees. It is proof against machine-gun fire up to 0.50 cal, but a 20 mm (0.8 in) cannon or anything heavier will penetrate and destroy the vehicle.

Driver-mechanic
He is in charge of the BMP when the infantry section and the commander dismount. In the Russian Army he is the equivalent of a lance-corporal and has had six months' specialist training.

able of offering the infantry at least some degree of protection, but none of these was capable of keeping up with and providing protection to fast-moving tanks. It became obvious that only tracks could keep up with tracks, and it was from this realisation that the BMP evolved.

Firepower

The BMP has considerable potential firepower. The ultra-low one-man turret mounts a 2A28 73-mm low-pressure smoothbore gun firing fin-stabilised HEAT (High-Explosive Anti-Tank) rounds, the automatic loader giving a firing rate of eight rounds per minute.

Storage of 73-mm ammunition is a problem. Thirty rounds are carried beneath the turret to the right of the gunner; not only does this make reloading difficult – the gunner must dodge the breech mechanism as it recoils, or suffer serious injury – but it means that any enemy anti-tank round that manages to penetrate the centre of the BMP's hull will almost certainly come into contact with the unused rounds, causing an explosion lethal to all inside.

The AT-3 'Sagger' anti-tank missile can be fired from a stubby rail immediately above the 73-mm gun barrel. Theoretically the BMP travels with one missile loaded on the rail, with a further three stored internally. There have, however, been numerous cases of the missile falling off the rail during rough cross-country training, and so it is far more likely that 'Sagger' will be loaded only at the last possible moment.

Unfortunately for the crew, 'Sagger' can only be loaded externally, thus making it necessary for a crew member to leave the protection of the interior to reload – not only dangerous, but uncomfortable for the other passengers if the vehicle is operating in an NBC environment. The protective seal will be broken, and everyone will have to put on hot and cumbersome rubber NBC suits.

All-round manoeuvre

In place of 'Sagger' a few BMPs mount the AT-4 'Spigot' missile on a portable mount on the turret. This missile, which bears a striking resemblance to the NATO MILAN, has an effective range of over 2,000 m (6,560 ft) against all but the heaviest armour.

The BMP's model V-6, six-cylinder in-line water-cooled engine develops 280 hp at 2,000 rpm and is situated to the right of the commander and to the rear of the driver, thus giving the passengers some degree of protection against frontal hits. It is powerful enough to give a fully-loaded BMP a top speed of 55 km/h (34 mph) on road and 40 km/h (25 mph) over deep snow, with a maximum range of 500 km (310 mph). It is, however, extremely noisy and vibrates uncomfortably at high speed, so much so that the gyroscopic navigation system becomes unreliable and has to be reset every 30 minutes.

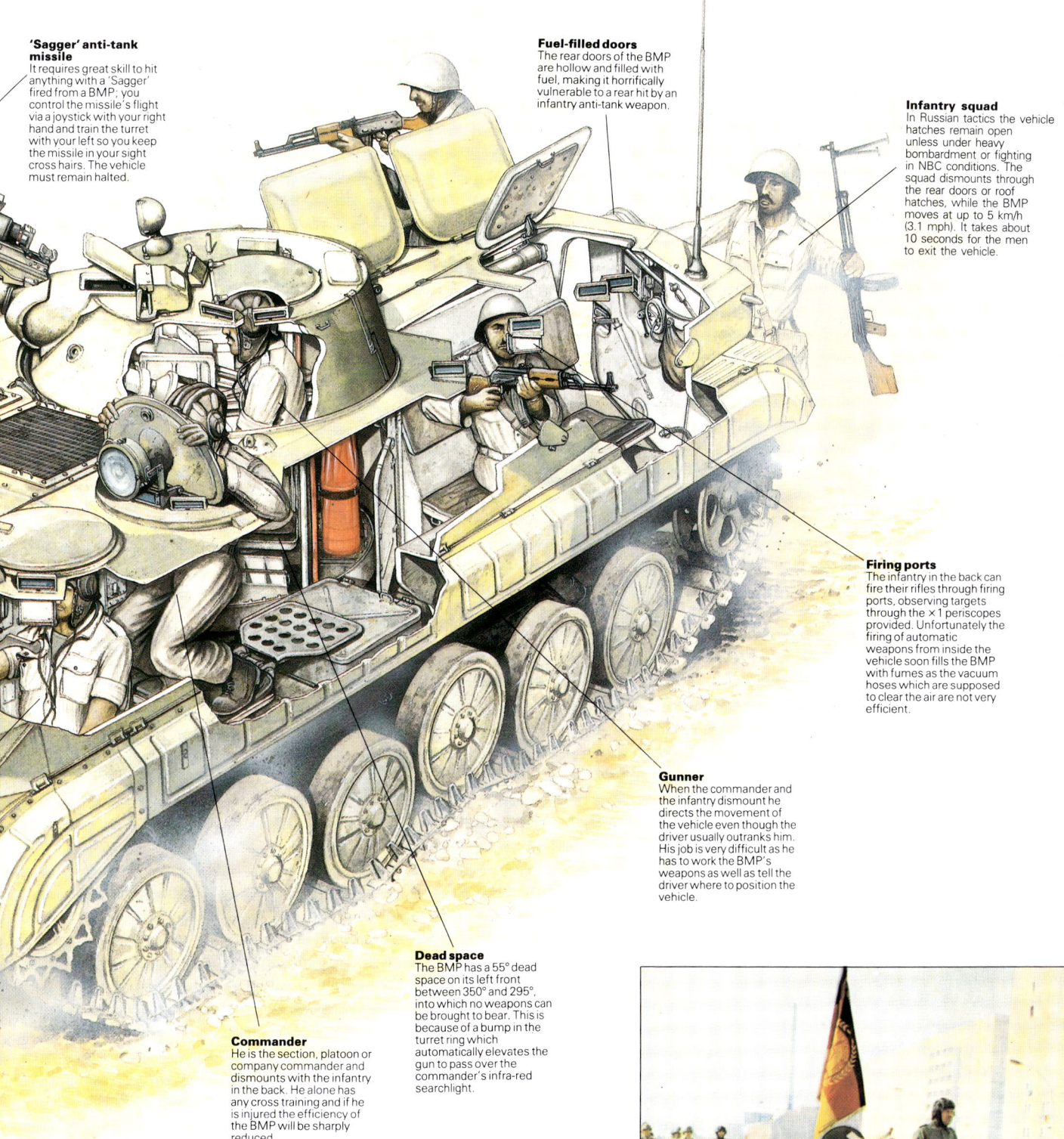

MECHANISED COMBAT

'Sagger' anti-tank missile
It requires great skill to hit anything with a 'Sagger' fired from a BMP; you control the missile's flight via a joystick with your right hand and train the turret with your left so you keep the missile in your sight cross hairs. The vehicle must remain halted.

Fuel-filled doors
The rear doors of the BMP are hollow and filled with fuel, making it horrifically vulnerable to a rear hit by an infantry anti-tank weapon.

Infantry squad
In Russian tactics the vehicle hatches remain open unless under heavy bombardment or fighting in NBC conditions. The squad dismounts through the rear doors or roof hatches, while the BMP moves at up to 5 km/h (3.1 mph). It takes about 10 seconds for the men to exit the vehicle.

Firing ports
The infantry in the back can fire their rifles through firing ports, observing targets through the ×1 periscopes provided. Unfortunately the firing of automatic weapons from inside the vehicle soon fills the BMP with fumes as the vacuum hoses which are supposed to clear the air are not very efficient.

Gunner
When the commander and the infantry dismount he directs the movement of the vehicle even though the driver usually outranks him. His job is very difficult as he has to work the BMP's weapons as well as tell the driver where to position the vehicle.

Dead space
The BMP has a 55° dead space on its left front between 350° and 295°, into which no weapons can be brought to bear. This is because of a bump in the turret ring which automatically elevates the gun to pass over the commander's infra-red searchlight.

Commander
He is the section, platoon or company commander and dismounts with the infantry in the back. He alone has any cross training and if he is injured the efficiency of the BMP will be sharply reduced.

The BMP is fully amphibious and can reach speeds of 8 km/h (4.9 mph) in water. But it cannot float very well, and although in theory it can climb gradients of 60 per cent, it is known to have difficulty negotiating gently sloping but slippery river banks. Before the BMP enters the water a telescopic schnorkel is raised behind the turret to provide air to the crew and passenger compartments, and a splash board is raised at the bow to give the driver some protection from icy waters. Despite this, river crossings must be very unnerving for the

A pre-unification BMP-1 heads for a parade with the commander standing up. In combat, the BMP's commander dismounts with the infantry section, and in the Russian army he is the only man in the section who knows how to use the radio or how to read a map.

155

MECHANISED COMBAT

crew and even more so for the passengers, who travel virtually submerged with little chance of escape in the event of an accident.

Although the commander and crew must exit from hatches on top of the hull, the passengers have the option of using either the four small top hatches or two rear doors. Although a good idea on the face of it, these doors have often proved a death trap: the doors have been fitted with large, lightly armoured fuel tanks to increase the range, and as the Israelis were soon to discover in the Yom Kippur War, a single incendiary round fired into the rear doors of a BMP invariably proves enough to set the vehicle ablaze.

NBC protection

The Russians treat NBC warfare very seriously, providing all BMPs with a protective system. An air intake immediately behind the turret draws contaminated air into a blower/dust separator. The air is then filtered and forced into the troop and driver's compartment through 11 outlet vents.

The forced air creates a build-up of pressure inside the vehicle, thus preventing NBC agents from leaking through such areas as the firing ports, which do not have airtight seals. Once a hatch is opened, for instance to load a fresh 'Sagger' missile, the vehicle must be decontaminated.

Several variants of the highly successful basic BMP have been introduced in the last 20 years, and many can still be seen in active service. The earliest variant, designated BMP-1K, is issued to company commanders and will normally be found behind the lead platoon.

From a distance the BMP-1K looks similar to the conventional model, but the side firing ports have been sealed

The rear doors of the BMP are its Achilles heel, filled with fuel and lightly armoured. A single hit here can incinerate the whole crew.

and extra antennae have been fitted.

Taking the basic BMP chassis, the Russians have added a two-man turret to create a new subfamily of surveillance vehicles. Artillery forward observation units are now issued with the BMP M1975, in which the 73-mm gun has been replaced by a 7.62-mm machine-gun and 'Small Feed' ground surveillance radar mounted on the rear of the turret. Commanders of Soviet divisional and regimental reconnaissance units are equipped with the BMP M1976, which retains the 73-mm gun (but not the 'Sagger' rail) in the two-man turret.

Improvements

The much-improved BMP-2, with its two-man turret and 30-mm gun, was first seen in public in 1982, by which time it had already deployed to Afghanistan and to Soviet forces in Germany. The BMP-2 is usually fitted with an AT-5 'Spandrel' anti-tank missile in place of the 'Sagger' fitted to earlier examples.

The latest BMP variant to have been revealed is the BMP-3, which has a large-calibre soft-recoil gun and a coaxially-mounted cannon. Its primary function is thought to be direct fire support for more conventional BMPs.

The BMP has been one of the greatest successes of Russian post-war military technology, and, even now, some 35 years after its introduction, it remains one of the most widely used mechanised infantry combat vehicles.

BMP-1 and BMP-2

BMP-1

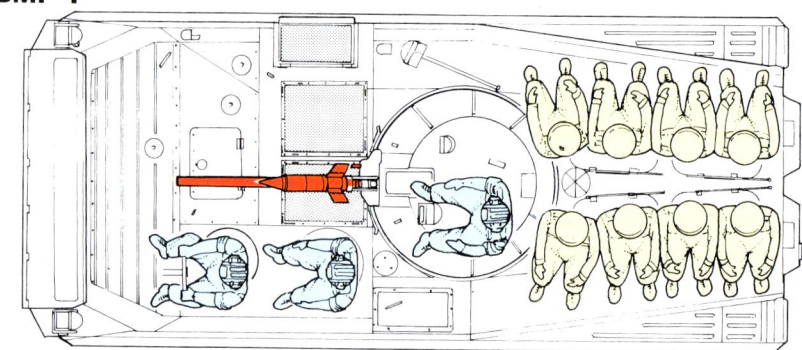

BMP-2

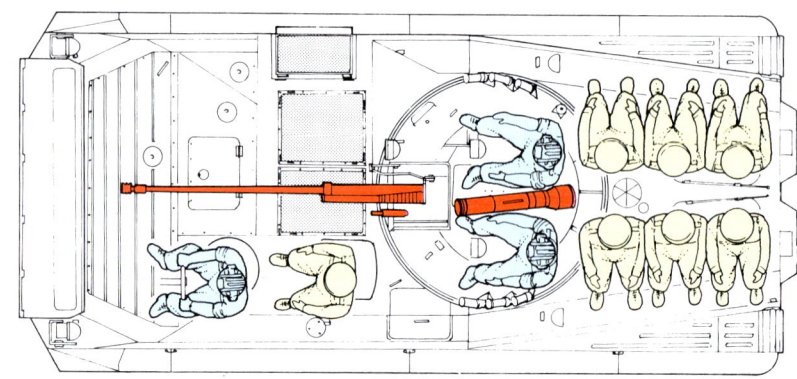

The BMP-2 appeared 20 years ago and is a substantial improvement, although the seating is reduced and the infantry section split up. The low-velocity 73-mm gun on the BMP-1 is better than the 30-mm cannon for taking out enemy infantry positions and will penetrate more armour, but its inaccuracy over 800 m (2,620 ft) is a serious handicap.

MECHANISED COMBAT

AMX-10P

ARMOURED PERSONNEL CARRIER

The AMX-10P is the replacement for the AMX-VCI, and was designed by the Atelier de Construction d'Issy-les-Moulineaux in the mid-1960s. Production was undertaken from 1972 at the Atelier de Construction Roanne (ARE), where manufacture of the AMX-30 MBT is undertaken together with that of the AMX-10RC (6×6) reconnaissance vehicle which is automotively related to the AMX-10P even though it is a wheeled vehicle. First production vehicles were completed in 1973, and since then more than 2,000 vehicles have been completed for the French army and for export to countries such as Greece, Indonesia, Mexico, Qatar, Saudi Arabia and the United Arab Emirates. The Saudi Arabian order was the largest export deal, with 300 AMX-10s purchased. The Saudi Arabian army had already bought AMX-30 tanks from France.

Aluminium hull
The AMX-10P has a hull of all-welded aluminium, with the driver at the front left, engine to his right, two-man turret in the centre and troop compartment at the rear. The eight troops enter and leave the vehicle via a power-operated ramp in the hull

Above: French army AMX-10P armoured personnel carriers race towards 'enemy' positions during an exercise. Note the well-sloped glacis.

Below: The AMX-10P was introduced to replace the AMX-VCI tracked personnel carrier based on the hull of the AMX-13 light tank seen here on parade in Paris.

MECHANISED COMBAT

The AMX-10P has a reasonable cross-country performance for a 1960s APC design. It is capable of a maximum road speed of 65 km/h (40.6 mph) and exerts a ground pressure of 0.53 km² (0.2 m²).

rear; there is a two-part roof hatch above the troop compartment. Apart from the roof hatches and two firing ports in the ramp, there is no provision for the troops to use their rifles from within the vehicle. The power-operated turret is armed with a 20-mm dual-feed (HE and AP ammunition) cannon with a co-axial 7.62-mm machine-gun; mounted on each side of the turret are two smoke dischargers. The weapons have an elevation of +50 degrees and a depression of −8 degrees, turret traverse being 360 degrees. Totals of 800 rounds of 20-mm and 2,000 rounds of 7.62-mm ammunition are carried.

The AMX-10P is fully amphibious, being propelled in the water by waterjets at the rear of the hull, and is also fitted with an NBC system and a full range of night vision equipment for the commander, gunner and driver.

AMX-10 variants

Variants of the AMX-10P include an ambulance, a driver training vehicle, a repair vehicle with crane for lifting engines and an anti-tank vehicle with four HOT ATGW in the ready-to-launch position. Another 14 missiles are carried inside the vehicle at the hull rear. The crew of five consists of the commander and gunner, who are stationed in the 'Lancelot' turret, plus two missile loaders and the driver. The turret has power traverse with manual back-up. The gunner is seated on the left of the turret and he observes through an M509 sight with ×3 magnification for acquiring targets and ×12 magnification for engaging them. The commander has a laser rangefinder with a maximum range of 8,000 m (26,246 ft), plus optical sights of x3.5 and x8 magnification. The combat weight is 14.1 tonnes (13.8 tons). The AMX-10 HOT has not been adopted by the French army and the only export customer to date is the Saudi Arabian army.

Other AMX-10 variants include the AMX-10PC command vehicle, a RATAC radar vehicle, artillery observation and fire control vehicles, an AMX-10TM mortar tractor towing 120-mm Brandt mortar and carrying 60 mortar bombs and the 81-mm fire-support vehicle fitted with the Thomson-Brandt 81-mm smoothbore gun. The AMX-10 PAC 90 fire-support vehicle is another version developed for the export market. First announced in 1978 the PAC 90 has a standard AMX-10P hull fitted with the GIAT TS90 turret – the turret also mounted on the Panhard ERC 90 F4

AMX-10P
Armoured personnel carrier

Developed during the mid-1960s, the AMX-10P is the French army's equivalent of the British Warrior. It is lightly armoured but equipped with a 20-mm cannon capable of destroying similar APCs. Fully amphibious, it has a crew of three and carries an eight-man infantry section in the troop compartment.

MECHANISED COMBAT

Above and above right: French infantry de-bus at speed from the crew compartment. Note their short and handy FAMAS 5.56-mm rifles, well-suited to this role.

Toucan II turret
French army AMX-10Ps are fitted with the Toucan II turret, with the gunner on the left and the commander seated on the right. It is fitted with a 20-mm cannon, a co-axial 7.62-mm machine-gun and a searchlight.

Troop compartment
The AMX-10P has a ramp in the hull rear rather like the US M113. This allows the eight-man infantry section to deploy as fast as possible.

NBC system
Fitted to the right side of the hull, the NBC unit filters out contaminated air to enable the AMX-10P to remain operational on the nuclear battlefield.

Torsion bar suspension
The AMX-10P has five road wheels, with the drive sprocket at the front and the idler at the rear. There are hydraulic shock absorbers on the first and fifth wheel stations. The tracks are fitted with replaceable rubber pads.

MECHANISED COMBAT

donesia also took delivery of a number of AMX-10Ps with the original two-man turret replaced by a new one-man turret at the rear armed with a 12.7-mm M2 HB machine-gun. These vehicles were all modified for amphibious operations. They have 305-mm water-jets that propel them in the water at up to 10 km/h (6.25 mph). The waterproofing is reinforced and the vehicle has special protection against salt corrosion. A transparent wavebreaker is mounted on the hull front, which can be raised or lowered hydraulically by the driver. The AMX-10P has four bilge pumps, different air intakes and a pneumatic emergency engine starter. When 'swimming' air is drawn into the hull via the circular roof cowling and the hot air is expelled through side vents.

Left: Although there are many versions of the AMX-10P in service, the French army uses the VAB wheeled armoured personnel carrier alongside them. Cheaper to build and operate, its cross-country performance is only marginally inferior. This is the Mephisto – the anti-tank variant fitted with HOT missiles.

Right: The 14.2-tonne (14-ton) AMX-10P is fully amphibious. The trim vane can be seen here in the lowered position on the hull front.

Sagaie and the Renault VBC 90 6×6 armoured cars.

The PAC 90 is designed for anti-tank missions as well as fire support for infantry. It could also serve as a reconnaissance vehicle or APC. The 90-mm gun fires canister, HE, HEAT, smoke and APFSDS. The normal ammunition load is 12 HEAT and eight HE carried in the turret, 16 in the turret bustle behind the commander and four more behind the commander's seat. In addition to commanding the vehicle, the commander has to work as the loader.

Indonesian service

In addition to its three-man crew of commander, gunner and driver, the AMX-10 PAC 90 carries four infantrymen in the rear. The vehicles delivered to Indonesia have improved amphibious characteristics and they are meant to leave landing craft offshore rather than just cross rivers and streams, as is the basic vehicle. In-

Since World War I, French armoured vehicles have sometimes carried the names of battles. This AMX-10P 'Jena' is named after Napoleon's 1806 victory over the Prussians. The vehicle is presumably not part of the Franco-German brigade.

MECHANISED COMBAT

ON THE BORDER WITH THE BUFFEL

The Buffel is a highly-effective mine-protected fighting vehicle used by the South African Defence Force. It has been used on operations for several years, proving itself time and time again in hostile environments, and was also exported to Sri Lanka for use against the Tamil guerrillas. Manufacture of the Buffel is controlled by Armscor, South Africa's enormous defence production conglomerate. Vigorous trials were carried out in the unforgiving terrain of Ovamboland in north-east South West Africa (Namibia).

Its mine protection characteristics are invaluable in the mine-infested SWA/Angolan border areas, and its powerful engine and cross-country capabilities enable it to negotiate the rugged bush and 'shonas' (river beds) of the area. It incorporates a number of successful ideas from the highly effective vehicles used by the Zimbabwean army in its counter-insurgency (COIN) operations.

The Buffel is designed to provide mobility, mine/bullet/shrapnel protection and enhanced fighting capability to COIN forces during rural operations. It has a long operating range (580 km [360 miles] cross-country), good cross-country ability, average protection and good combat ability.

The South African Defence Forces have three types of mine-resistant vehicle. MPFVs (mine-protected fighting vehicles), for instance the Buffel and Ratel, are designed to protect their occupants and operate in an offensive role. MPVs (mine-protected vehicles) are intended for the defensive role, purely to provide protection for the occupants against mine explosions (for instance, the Hippo, Roibok and Zebra). AMVs (anti-mine vehicles) like the Spinnekop are designed to locate or detonate mines and are the

A Buffel mine-protected vehicle speeds across the seemingly limitless expanse of the Namibian desert. In such conditions, a vehicle's range is one of its most important attributes.

MECHANISED COMBAT

Inside the Buffel

In the 1970s South African and Rhodesian forces developed a series of novel APCs to supplement their stock of ageing British vehicles. Many of these vehicles were improvised conversions based on what was available, e.g. Unimog truck chassis. The Buffel was the first to be manufactured in large numbers. It is a useful, long-range APC for COIN operations.

The Buffel pays a price for its ability to survive mines: its centre of gravity is uncomfortably high, which makes it teeter alarmingly during a tight turn and over rough ground. It is a vehicle that certainly takes some getting used to.

Powerplant
The Buffel is powered by a 5.675-l (1.25-gal) Mercedez-Benz engine. Its maximum power is 92 kW at 2,800 rpm. Top speed on roads is 98 km/h (61 mph); cross-country 45 km/h (28 mph).

Protective grille
The Buffel is designed to crash through the bush at high speed in pursuit of elusive guerrillas. The vehicle and the crew face as much of a battle with the vegetation as their human enemies: Buffels have lost wheels to the thick roots encountered in northern Namibia.

most unusual-looking of the mine-resistant vehicles.

The Buffel carries an infantry section of 10 men and a driver, with their basic equipment and webbing. Extra equipment can be towed in a trailer. The section commander sits at the left front so that he can speak to the driver, operate the section grenade launcher, and control the front light machine-gun. The section's machine-gun team is at the left rear, with the No. 1 manning the weapon. The section 2IC controls the rear machine-gun. The remaining riflemen sit in sequence, with the No. 1 rifleman manning the front machine-gun: this MG is fitted to the vehicle and is not one of the section weapons. Some Buffels have floor clips for a 60-mm mortar, and if a mortar is mounted the mortar No. 1 and No. 2 sit opposite the tube.

Bulletproof glass

The vehicle has three basic components: the chassis and powerpack, the driver's cab, and the troop compartment. The cab is heavily protected and has bulletproof glass on all sides. The troop compartment is a strengthened steel box with an open-topped wedge-shaped base, secured to the chassis by cables.

The Powerplant consists of a 5.675-litre Mercedes Benz water-cooled, six-cylinder inline four-stroke direct-injection engine. This transmits its drive to the road wheels via a single dry-plate clutch and a transmission assembly. The transmission assembly provides eight forward and four reverse gears, as well as selectable four-wheel drive and lateral differential locks on both axles.

The Buffel can be employed offensively as a combat vehicle, but the extent to which it is deployed in this role depends on the terrain and the danger from enemy anti-tank weapons. Detailed drills are not laid down for mounted attacks, because tactics will be influenced by the ground and the tactical situation. The commander on the spot will be in the best position to determine the mode of action.

Shock action

The speed and firepower of the Buffel ensure surprise and shock action, and the protection provided against small-arms fire and shrapnel will minimise personnel losses. But the driver must remember that, although speed is an advantage for shock effect and for the momentum necessary to flatten obstructions, the Buffel is rather top heavy and lurches alarmingly during a tight turn. If he decides to make a 90-degree turn within seconds of commencing the assault, he will turn the vehicle over. Also, if he is stopped short of the objective by a large tree he will cause shock-effect to his crew but not to the enemy!

MECHANISED COMBAT

Roll bar

Troop compartment
When a Buffel is smashing its way through the bush, all men in the troop compartment must sit in their seats, strapped in, facing inwards. If not tightly secured you would be thrown around and seriously injured if the vehicle did hit a mine.

Driver's cab
Protected by bulletproof glass, the driver sits in an armoured box that is protected against small arms fire.

'V' shaped mine resistant hull
The Buffel is primarily designated to protect its occupants against mines. Open-topped and lightly armoured, it is vulnerable to artillery and cannon.

Rear suspension
This is a double-coil spiral with double-acting hydraulic telescopic shock absorber and stabiliser bar with torsion bar at the rear.

Front suspension
This consists of a single-coil spiral spring with double-acting hydraulic telescopic shock absorber and stabiliser bar.

Left: A quick glance at the Buffel shows its vulnerability to a hand-held anti-tank rocket such as the RPG-7. Only careful vigilance and a heavy volume of suppressive fire at the ready will protect the crew from a determined RPG gunner.

Right: A profile view shows the Buffel's impressive ground clearance, which allows drivers to bounce across the veldt to their heart's content. The troops in the back have to be strapped in anyway, according to regulations, to survive a mine blast.

MECHANISED COMBAT

In the event of a mine detonation the troops inside are protected by being securely strapped onto hard seats, and the open roof minimises the damage caused by overpressure. The wedge shape of the passenger compartment helps to deflect the force of an explosion away from the vehicle. If the vehicle is tipped over, protection is provided by a roll bar running the length of the compartment.

The water supply for the section (100 l [22 gal]) is in an area under the floor, which helps disperse the heat of

Below: The Buffel might look bizarre, but it was designed for counter-insurgency warfare. It gives security forces long-range mobility, and protection against mines.

Above: The Buffel has spawned a host of variations using the same chassis but different cab and troop compartments. This option at least keeps the snakes from falling into the back.

a mine blast. The wheels can also be filled with water to dampen blast: water is an excellent absorber of heat, and the intense heat generated in an explosion, almost 3,000°C (5,432°F), can be reduced by 50 per cent.

The occupants must wear their pilot-type safety harnesses at all times. Belts must be tight, otherwise whiplash injuries will occur. The manner in which occupants are sitting during an explosion is of great importance, too: the shock wave is absorbed far better if the spine is straight and supported with the feet flat on the floor. If you have your feet in the air during an explosion they will be forced towards the floor at a speed comparable to that of jumping off a three-storey building – with smashed legs as a result. And equipment and weapons must be fastened down or locked away – they become dangerous projectiles in an explosion.

South West Africa

Operating in the bush of northern South West Africa, the Buffel crashes through the low trees with branches dipping into the open troop compartment. You must keep your arms inside the vehicle and be ready to duck as the branches are very sharp and can do a lot of damage. After an hour or so going cross-country through the bush, the bottom of the troop compartment can fill up with twigs and leaves.

Even more embarrassing is the tendency of local snakes to bask on the lower branches: swept from their resting place and deposited at your feet, they usually suffer a sense of humour failure and bite. You, of course, are tightly strapped in with little chance of evasive action. Some commanders reckon they suffered more casualties from snake bites than from SWAPO.

The Buffel is a highly successful APC for long-range counter-insurgency operations. It has its limitations: it carries nothing heavier than small arms and its armour will not keep out modern anti-tank rockets, but its intended enemy relies heavily on mines and rarely operates in large formations. The Buffel enables relatively small numbers of regular troops to dominate large areas of ground which is, after all, the essence of counter-insurgency.

A Buffel at rest during operations along the border with Angola. Many South African vehicles carry names painted on the hull, as the long columns of vehicles driving back over the border in 1988 bore witness.

MECHANISED COMBAT

M113

A US Army M113 ACAV (Armored Cavalry Assault Vehicle) gives a lift to some US Marines in the northern province of the Republic of Vietnam. They are hunting out North Vietnamese regulars following the Tet offensive of 1968.

IN ACTION

In the summer of 1961 Viet Cong units fighting in the Mekong Delta suffered a series of surprise defeats at the hands of government troops equipped with new armoured vehicles. These large olive-drab machines roared across the paddy fields emitting clouds of fumes and smoke and spraying the guerrillas with heavy machine-gun fire. To the US Army who had supplied them they were M113 armoured personnel carriers, but to the panic-stricken VC they were 'Green Dragons': sinister machines to be treated with respect.

While it did not take long for the Viet Cong to develop tactics to use against the M113, their shortage of anti-tank weapons made this simple APC a very effective weapon in the early years of the US involvement in South East Asia.

Jack-of-all trades

From its inception the M113 was designed to be a jack-of-all trades. The original US Army specification called for a lightweight, amphibious and air-droppable APC with good cross-country performance. By application of kits and modifications of the basic chassis a generation of support vehicles could be created, all sharing the same automotive parts. By the 1980s the M113 had become the most widely used armoured fighting vehicle in the world. Over 50 armies are equipped with it, and most have converted it to suit their own special requirements.

The M113 followed a number of US Army APCs, all of simple box-like design, for transporting the infantry into battle. The armour protection was designed to defeat small arms and shell splinters but nothing more; the infantry squad riding in the vehicle was supposed to dismount to assault an enemy position, while the APC provided supporting fire with its .50-cal machine-gun. However, as it received its combat debut in Vietnam, mounted action seemed to offer exciting possibilities.

Combat experience soon led to a number of alterations in the M113's design. Most significantly, the petrol

165

MECHANISED COMBAT

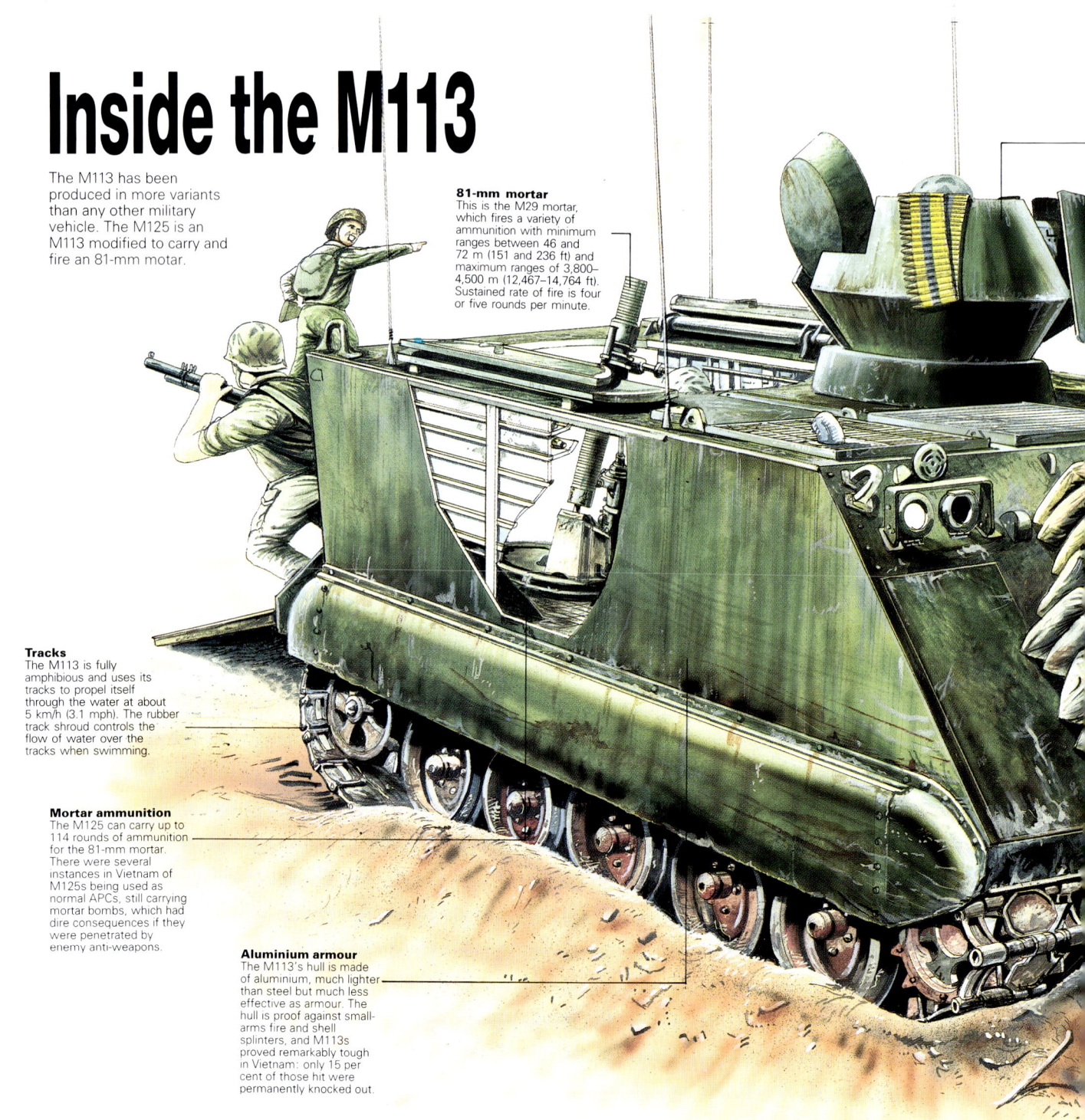

Inside the M113

The M113 has been produced in more variants than any other military vehicle. The M125 is an M113 modified to carry and fire an 81-mm motar.

81-mm mortar
This is the M29 mortar, which fires a variety of ammunition with minimum ranges between 46 and 72 m (151 and 236 ft) and maximum ranges of 3,800–4,500 m (12,467–14,764 ft). Sustained rate of fire is four or five rounds per minute.

Tracks
The M113 is fully amphibious and uses its tracks to propel itself through the water at about 5 km/h (3.1 mph). The rubber track shroud controls the flow of water over the tracks when swimming.

Mortar ammunition
The M125 can carry up to 114 rounds of ammunition for the 81-mm mortar. There were several instances in Vietnam of M125s being used as normal APCs, still carrying mortar bombs, which had dire consequences if they were penetrated by enemy anti-weapons.

Aluminium armour
The M113's hull is made of aluminium, much lighter than steel but much less effective as armour. The hull is proof against small-arms fire and shell splinters, and M113s proved remarkably tough in Vietnam: only 15 per cent of those hit were permanently knocked out.

engine, which was a serious fire hazard in a damaged vehicle, was replaced by a diesel. Unlike petrol, diesel fuel will not be ignited by a penetrating anti-tank round.

M113s in Vietnam soon began to appear very different from the gleaming vehicles equipping US units in Europe; the APC's uncluttered box shape vanished beneath a pile of sandbags, ammunition boxes and spare track blocks, all placed to detonate a

B Company, 1/50th Infantry, 173rd Airborne Brigade advance on the village of Phu Loc during a search for local Viet Cong and two enemy nurses in May 1968. Unusually, the M113s still carry prominent US Army markings.

MECHANISED COMBAT

Additional armour
The South Vietnamese experience with M113s in the early 1960s revealed that the machine-gunner was an easy target for enemy infantrymen, and M113s soon sprouted extra armour plate. The back plate is a two-part section overlapped by the large gunshield fitted to the machine-gun.

.50-cal machine-gun
The M125 retains the standard machine gun armament of the M113 APC. This was essential in Vietnam, where convoys were frequently ambushed at very short range.

Driver
The basic M113 is driven with conventional steering brake levers, although the German firm of Thyssen Henschel offers a steering modernisation package which substantially improves its handling.

Splash plate
Lowered for amphibious operation, the splash plate was often dropped and piled full of sandbags to provide a high degree of protection against RPG-7 anti-tank rockets.

Engine compartment
The first production M113s had a petrol engine, which is a significant fire risk. From the M113A1, all models have been fitted with a GMD Detroit Diesel six-cylinder water-cooled diesel developing 215 hp at 2800 rpm.

The M577 is an M113 with a much higher roof, used as a mobile command post or a medical treatment vehicle. It can carry an external generator to power its communications equipment.

power, despite the introduction of the TOW-armed Bradley. The US Army has some 1,400 M113s fitted with TOW anti-tank guided missiles which are launched from a pop-up pedestal that retracts into the troop compartment. The M901 'Improved TOW' vehicle is simply a more capable anti-tank variant with the Emerson ground-launched TOW system in an armoured launcher, and a target acquisition sight mounted on two arms. Currently, the US Army has over 500 M901s in service.

Less exciting to watch, but of enormous tactical importance, is the M577 series of command post vehicles. By raising the roof behind the driver enough space is created in the troop compartment for a command team to have room to operate; more space can be created by fitting a tent to the back when the vehicle is in a static position. An externally mounted generator provides power for extra radios and associated command equipment. This model of the M113 is also suitable for mobile medical treatment.

To dig fellow members of the M113 family out of trouble, the US Army developed the M113A2 recovery vehicle. Its hydraulic winch and auxiliary crane enable it to recover vehicles after it has dug spades into the ground to anchor itself firmly. This is a far cry from the Vietnam days, when all manner of ingenious devices were invented to allow basic M113s to trundle around the waterlogged Mekong Delta. Block and tackle systems were fitted to drag APCs over terrain that was too wet to drive over but too dry to use the M113's amphibious ability.

Daisy chain

Thanks to its relatively low ground pressure, the M113 manages fairly well in boggy ground if enough operate together. In the Mekong Delta the 'Daisy Chain' method of linking up 15 APCs together enabled rice paddies to be crossed successfully.

More recent additions to the M113 family include the M1059 smoke gen-

shaped-charge anti-tank round before it struck the armour and thus reduce its ability to penetrate.

The M113's armament of one .50-cal machine-gun was soon supplemented in Vietnam. Christened the Armored Cavalry Assault Vehicle, its machine-gun protected by a gunshield and back plate and with a pair of 7.62-mm M60s or Browning .30 cals fitted to the sides firing from the crew compartment, the M113 went to war. Subsequently the vehicle has been armed with 81-mm and 107-mm mortars, anti-tank guided missiles, flamethrowers, 20-mm cannon, Vulcan six-barrelled anti-aircraft Gatling guns, and even fitted with turrets for tank guns.

Incredibly versatile

In other support roles the M113 continues to be incredibly versatile. It provides all the US Army's mobile air defence: after the humiliating failure of the Sergeant York 40-mm anti-aircraft gun system, the M168 Vulcan soldiers on. This is an M113 fitted with the Navy-modified 20-mm gun with a gyro lead computing sight and a range-determining radar attached. The Army's self-propelled Chapparal surface-to-air missile system is mounted on the M548 tracked cargo carrier chassis, another development of the ubiquitous M113.

M113s will continue to provide considerable long-range anti-tank fire-

MECHANISED COMBAT

US Army M113s gather in Germany in the mid-1980s. Although supplanted by the larger and heavier M2 Bradley, the M113 remains a highly mobile APC.

erator, a self-propelled and self-explanatory vehicle of which nearly 200 will soon be in US service. FMC have developed the original vehicle out of all recognition and produced the 'Armoured Infantry Fighting Vehicle', an MICV of strikingly modern appearance and high capability, yet costing a fraction of the bill for an M2 Bradley. This is a private venture calculated to win export sales and place another question mark on the Bradley's cost-effectiveness.

Like most NATO forces, the German army uses the M113, and Thyssen Henschel have developed a much improved model with modern steering gear. A steering wheel and footbrake replace the traditional steering brake levers and offer many advantages: direction-keeping is more stable through continuous steering and manoeuvrability is greatly increased. The Australian army operates 46 M113s fitted with the turret of the British Scorpion reconnaissance vehicle armed with a 76-mm gun. These replace 18 M113s which carried the turret of the old Saladin armoured car. Standard Australian M113s carry a turret with a .30-cal machine-gun and 5,000 rounds of ammunition.

Zeldas

The Israeli army has made wide use of M113s since the late 1960s, when they began to replace the World War II half-tracks still in service with the IDF. When the Israelis invaded Lebanon in 1982 their M113s, called Zeldas, were protected by a thick extra layer of appliqué armour, which helps protect them against the HEAT warheads of RPG-7 rocket launchers, the most widely encountered anti-tank weapons in the Middle East.

Over 40 years after its introduction the M113 is more popular than ever. The US army continues to modernise its vast fleet of M113s, updating powerplant, transmission and performance. New models with stretched crew compartments and various armaments are offered for export, and the Army is studying ways of fitting both active and passive NBC systems and appliqué armour along the lines of the Israeli model.

In Vietnam, where the M113 received its baptism of fire, thousands of vehicles supplied to the South were captured in 1975 and a few 'Green Dragons' continue to serve their new masters.

The M113 has been used to mount a variety of weapons systems. This is the M163 self-propelled AA gun system, in operation in Saudi Arabia just before the start of the Gulf War of 1991.

MECHANISED COMBAT

Since its introduction in the 1960s, the Commando has been the inspiration for a wide range of wheeled armoured vehicles, ranging right up to the powerful six-wheeled V-600 armoured reconnaissance vehicle.

COMMANDO!

Left: The original V-100 Commando was developed as a private venture in the early 1960s. It was used extensively by the US forces in South East Asia, who needed a modern vehicle for a variety of tasks, ranging from convoy escort to air base perimeter security.

The Commando multi-mission vehicle series is perhaps the most successful of its kind ever to have been produced. Within 10 years of completion of the first prototype in 1963, over 2,000 vehicles had been delivered to more than 15 countries throughout the world, and numerous variants had been designed to meet special requirements in the Third World. The Commando saw extensive service as an escort with the US Army, US Air Force and the army of the Republic of Vietnam between 1965 and 1975. It proved a reliable and popular vehicle, and today can be seen patrolling borders and flash points in Africa and the Middle East.

The original V-100 was followed by the larger V-200, and in 1971 by the V-150; the latter was the most successful variant of all. The V-300 was developed in the late 1970s to meet a demand for an increased payload, and was followed by the V-600 6×6 armoured car in 1985. It is interesting to note that the manufacturers, Cadillac Gage Corporation of Warren, Michigan, developed the V-600 as an entirely private venture in the optimistic belief that there will always be a market for wheeled light tanks.

169

MECHANISED COMBAT

The superstructure

The following specifically describes the V-150 Commando, but all except the latest models are similar in basic design. Although the V-150 is classed as an armoured car, it can carry 12 fully-equipped infantrymen in reasonable comfort.

The all-welded hull affords the crew and passengers reasonable protection against indirect small arms fire, but will not withstand a high-velocity 7.62-mm round striking at right angles or give much protection against larger machine-guns or shrapnel. The V-150 would be at a distinct disadvantage taking on a faster and better-armoured vehicle such as the British Fox.

Driver's position

The driver sits to the left of the commander in a small compartment to the front of the upper hull. When conditions allow, he drives with his head exposed through the double upper hatch, which opens outwards to the left and right. When forced to drive closed down, the driver relies on a series of five vision blocks arranged at the front and sides of his compartment.

There is a passenger hatch in each side of the hull about half-way along the main superstructure. The top opens sideways and to the rear to create maximum visibility for passengers and to allow a further machine-gun to be mounted if need be. The bottom half folds downwards, to form a step. Easy de-bussing is crucial in an ambush, when infantry need to deploy from the vehicle and counter-attack within seconds of the initial contact. A further hatch, situated in the hull rear on the right, opens upwards and downwards.

Each hatch has a vision block and firing point. A further six firing points, three on each side of the hull, allow for all-round small arms fire in support of the main armament. There is also a roof hatch to the right of the engine compartment. However, all small arms emit gases from the breech. After using the gun ports, even advanced APCs, such as the Soviet BMP-1 with its internal forced-air system, fill unpleasantly with these excess fumes.

Powerpack

The V-150 is fitted with either the Chrysler 361 V-8 petrol engine developing 200 hp or the Cummings V-6 diesel developing 155 hp. Engine and transmission are at the rear left of the hull, with easy access via a series of hatches on the roof and sides. A manual gear box with five forward and one reverse gear is normally fit-

The V-150 armed with a Cockerill 90-mm gun is now available with a Marconi digital fire control system, which significantly increases first-round hit probability.

The Commando Recovery Vehicle has a heavy duty winch with a maximum capacity of over 11 tonnes. Armament is usually the Browning .50-cal seen here, and 2200 rounds can be carried.

Oerlikon 204 GK 20-mm cannon
The V-150 can be fitted with either Oerlikon 20-mm cannon or M242 25-mm Chain Gun plus co-axial 7.62-mm machine-gun.

Powered turret
Fitted with Oerlikon 20-mm cannon, the turret tracks at 60 degrees per second. With the much heavier Chain Gun installed, this speed is halved, as is your ammunition supply.

Engine compartment
This is on the left-hand side of the hull rear and is accessed from the top and the side. The unit contains an integral fire extinguisher, manually operated by the driver.

Run-flat tyres
These are supposed to be good for another 40 km (25 mph) even when shot full of holes.

The Commando V-150S is 46 cm (18 in) longer than the V-150, giving more internal space and enabling the maximum load to be increased by 726 kg (1,600 lb) to 3,357 kg (7,400 lb). The longer wheelbase improves cross-country performance.

The V-150 mortar carrier is fitted with a US M29 81-mm mortar on a turntable with 360-degree traverse. Minimum range is 150 m (492 ft), maximum is 4,400 m (14,435 ft), and 62 rounds of ammunition is the normal load.

With a crew of three and room for two infantrymen, the V-150 with Oerlikon 20-mm cannon is a useful recce vehicle. A co-axial 7.62-mm machine-gun is fitted to the left of the cannon in addition to the AA machine-gun.

MECHANISED COMBAT

Inside the Commando

Seen here in US Army desert pattern, the V-150 series have been used in combat all over the world. The cannon-armed versions have much smaller troop-carrying capacity than the more lightly armed APC models.

Gunner
Either the gunner or the commander can fire the main armament. Rate of fire is either 1, 2 or 4 rounds per second or full auto (1000 rpm).

Driver

All-welded steel hull
This is proof against 7.62-mm small-arms fire, shell splinters and Molotov cocktails. All members of the Commando series are fully amphibious, being propelled in the water by their wheels.

Vision blocks
All vision blocks have crash pads to cushion your skull for when the moving vehicle lurches and you headbutt the side. Spall panels reduce the amount of splinter damage from non-penetrating rounds hitting the hull.

ted, but can be replaced by a fully automatic gearbox with three forward and one reverse gear if required.

Modified 5.08-tonne (5-ton) truck axles with locking differentials are fitted, as are run-flat tyres and power-assisted steering. A quite basic supension system of leaf springs and hydraulic shock absorbers gives the passengers an uncomfortable ride over rough tracks.

The Commando can rapidly be made fully amphibious by fitting two electric bilge pumps. It is propelled through the water by its wheels.

Maximum road speed is 88 km/h (55 mph), and maximum range 965 km (603 miles) on roadways and 680 km (425 miles) cross-country. Capable also of climbing gradients of 60 per cent, the V-150 is a highly versatile vehicle, ideally suited to the rough terrains in which it usually operates.

All models have a 4536-kg (9,980-lb) capacity winch fitted to the front of the hull. Like most vehicles designed primarily for export to the Third World, the Commando has neither an NBC system nor infra-red driving lights, although the latter can be fitted to order.

Firepower

The Commando can carry a wide variety of main weapons, from a single 7.62-mm machine-gun to a 90-mm cannon. One of two basic turrets can be mounted in the centre of the V-150's hull. The following are the basic variants available:

Armoured personnel carrier – with a pintle-mounted 7.62-mm or 0.3-in machine-gun with 3,200 rounds. The turret traverses through 360°, has maximum elevation of +59° and depression of −14°.

Twin combination – with a machine-gun turret and a one-man turret with a single-piece hatch opening to the rear. Two 0.3-in or 7.62-mm machine-guns can be mounted, or either of these can

Below: Philippine Marines, supporters of President Ferdinand Marcos, drive their V-150 Commando through the streets of Manila in an unsuccessful attempt to quell the unrest which was eventually to lead to the overthrow of the President.

171

MECHANISED COMBAT

be co-mounted with a 0.5-in heavy machine-gun. Ammunition stowage varies with the weapons fitted; a twin 7.62-mm installation would have 3,800 rounds. This model, which weighs 8,437 kg (18,600 lb), carries only seven infantrymen.

Mortar carrier – for a crew of five, and armed with an 81-mm mortar and 60 to 80 mortar bombs. Up to three 7.62-mm machine-guns and 2,000 rounds of ammunition can be carried.

20-mm turret model – fitted with a 20-mm Oerlikon cannon and co-axial 7.62-mm machine-gun. A Cadillac Gage electrohydraulic elevation/traverse mechanism is fitted to facilitate easy targeting. Total ammunition capacity is 400 rounds of 20-mm and 3,000 rounds of 7.62-mm ammunition. Smoke dischargers can be fitted to the turret rear if required. A crew of three plus five infantrymen can be carried.

76-mm turret model – fitted with the British 76-mm L23A1 gun and co-axial 0.3-in or 7.62-mm machine-gun with 41 and 3,600 rounds of ammunition, respectively.

90-mm turret model – with a 90-mm Mecar gun and co-axial 7.62-mm machine-gun. A further 7.62-mm anti-aircraft machine-gun can also be fitted to the roof. Forty-one rounds of 90-mm and 2,600 rounds of 7.62-mm are carried. By far the most potent variant, this can carry three infantrymen as well as the crew of three. It weighs 9,525 kg (21,100 lb) and is 2.54 m (8.4 in) high.

A number of support and paramilitary variants also exist, including a TOW anti-tank model with a single launcher and seven missiles, a command model with additional radio equipment and base security, and police emergency rescue models with fixed armoured pods on the roof, bi-fold doors and 13 gun ports.

A recovery vehicle with a 11,340 kg (25,000 lb) winch, 4,536 kg (10,000 lb) A-frame crane and 7.62-mm machine-gun completes the inventory.

The future

Over 3,000 V-100, V-150 and V-200 Commando multi-role vehicles have now been built by Cadillac Gage, the vast majority for export. Over 240 V-150s are now in service with the Saudi Arabian paramilitary National Guard alone.

The V-300 was developed in the late 1970s in response to requests for a heavy-duty 'big brother' for the V-150. Although it was unsuccessful in the United States' LAV (Light Armored Vehicle) evaluation, a number have nevertheless been built, including TOW anti-tank, command post, ambulance and 81-mm mortar versions. In mid-1988 sales had been made to Panama and Kuwait.

The 6×6 Commando V-600, unveiled in 1985, is the latest in the highly successful Commando series. It has a four-man crew and is fitted with the same Cadillac Gage turret as the Stingray light tank. Equipped with the British Royal Ordnance 105-mm Low Recoil Force Gun and the Marconi Command and Control Digital Fire-Control System, it is a match for any light tank in existence. Whether or not it will be adopted by the United States or West European nations remains to be seen.

Above: The V-300 Commando is a six-wheeled private-venture development based on the success of the V-150. It is in service in Panama and Kuwait.

Above: The basic Commando principles were established by the first V-100 in Vietnam: reliability and serviceability based on easily obtainable components.

Below: Commandos are not front-line vehicles, but they are more than sufficient for escort, base defence and internal security tasks.

MECHANISED COMBAT

THE BRADLEY IN ACTION

Seen here demonstrating its incredible speed cross-country, the Bradley is designed to fight in close conjunction with the M1 Abrams Main Battle Tank and is the most expensive and best-equipped vehicle of its kind.

The concept of the armoured personnel carrier is as old as that of the tank, and experience has shown that tanks need close support from their own infantry. This was repeatedly demonstrated in World War II and more recently in 1973 when Israeli tanks charged alone against Egyptian positions along the Suez Canal – and were bloodily repulsed. All mechanised armies employed APCs as battlefield taxis to allow their infantry to manoeuvre on to the battlefield while still protected against shell splinters and small-arms fire.

Building on their World War II experience, the Soviets went a step further. In the mid-1960s they introduced the BMP, a tracked APC fitted with a turret-mounted 73-mm low velocity gun and the AT-3 'Sagger' anti-tank guided missile. The infantry companies within Soviet armoured divisions now rode into battle in the BMP and were trained to fight from the vehicle as well as dismounted. With its gun to provide supporting fire and engage enemy APCs and a missile system able to knock out Main Battle Tanks, the BMP gave Soviet infantry a tremendous advantage.

Outperforming the rest

The Bradley is designed to trump the BMP. Its McDonnell Douglas 25-mm Chain Gun fires APDS (Armour-Piercing Discarding Sabot) shells at up to 500 rounds per minute. It is accurate at up to 2,500 m (8,202 ft), and its twin TOW anti-tank guided missiles can knock out any tank currently in service at over 3,000 m (9,840 ft). Expressly designed for speed and manoeuvrability, the Bradley handles well and its cross-country ability is substantially superior to that of the old M113.

The Americans had first experimented with Infantry Fighting Vehicles in Vietnam, where they fitted extra machine-guns to M113 Armored Cavalry Vehicles. The 'ACAVs' used fire-and-manoeuvre tactics to defeat the Viet Cong. ACAV crews fought from the vehicle instead of plunging into the undergrowth; they responded to ambushes by accelerating and driving into the enemy positions, firing in all directions. Where the enemy were short of effective anti-tank weapons this proved devastatingly effective – the Viet Cong had no answer to aggressively-handled APCs with awesome firepower.

Rifle firing port

In addition to the turret-mounted hardware, each infantryman inside the Bradley has a firing port through which he can fire an M231 5.56-mm rifle, essentially a cut-down M16 with an extendable wire stock which is de-

You exit the Bradley through a large, hydraulically-operating ramp in the rear of the hull.

173

MECHANISED COMBAT

The troop compartment of the Bradley is very cramped, but not such a tight squeeze as in the Russian BMP series IFVs. The Bradley has to be able to keep pace with the gas turbine-powered M1 tanks, but high-speed cross-country driving is very exhausting.

signed to be used from the cramped interior of a combat vehicle. With a cyclic rate of fire of between 1,100 and 1,300 rounds per minute, you can empty the 30-round magazine in under 1½ seconds! For this reason the Bradley carries 600 rounds per firing port weapon.

However, as the Bradley programme gathered pace, many criticisms emerged. It has been claimed that the Bradley's armour was not even proof against Russian machine-guns firing their new armour-piercing ammunition; that the Bradley was a step backwards from the M113, which carried a bigger infantry squad; and that it was incapable of keeping up with the gas turbine-powered M1 Abrams MBT. As for the opposition, the Russians had already updated the BMP by adding a new turret carrying a fully stabilised 30-mm cannon and an AT-5 'Spandrel' anti-tank missile-launcher. A special mount allows the 'Spandrel' missiles to be fired outside the vehicle.

Computer-tested

Under questioning, it was admitted that most of the tests that proved the Bradley could withstand hits from anti-tank weapons were computer simulations, not live firing tests. The US Army promptly organised a series of live firing tests to examine the performance of the Bradley's defensive systems, only for critics to claim these tests were rigged.
The Bradley was tested in two phases: Phase 1 against current Russian weapons known to be in service, and Phase 2 against projected weapons with an improved performance.

Fudges and fixes

The critics of the first tests cited a damning series of fudges and fixes used to pass the Bradley. Romanian-manufactured RPG-7 anti-tank rockets were fired from 18 m (59 ft), which is so close that the rocket does not have time to reach full velocity. TOW anti-tank missiles were exploded statically, and when one was

detonated at a 25° angle at the insistence of an Air Force Colonel overseeing the test, the result was catastrophic. Pressure and temperature effects were double the levels expected, and all hatches, including the heavy rear ramp, were consequently blown out.

It was claimed that in the initial tests, shots were deliberately aimed away from the vulnerable fuel and ammunition storage areas and that the dummies inside representing the crew members were soaked in water to stop their clothes catching fire. The Bradley has been fitted with an inert

Inside the Bradley

The M2 Bradley Infantry Fighting Vehicle is protected by all-welded aluminium armour with spaced laminate armour fitted to the hull, sides and rear. Its tremendous armament gives it the edge over any rival IFV, but doubts remain about its ability to survive a hit from a large-calibre weapon.

Turret
Part steel and part aluminium armoured, the turret has 360° traverse, moving at 60° per second, and can elevate the cannon and machine-gun to +60. The turret drive and stabilisation system allows the cannon to be accurately aimed and fired even when the vehicle is moving.

7.62-mm M240C co-axial machine-gun
This has 800 rounds at the ready and another 1,540 in reserve. It is mainly for anti-personnel use

McDonnell Douglas Helicopter Company M242 25-mm Chain Gun
The gunner sets the cannon to fire at either 100, 200 or 500 rounds per minute. All moving parts are operated by a single double-row roller chain which cycles in a racetrack pattern. Spent cases are ejected forward, out of the turret, and a dual feed system allows the gunner to switch ammunition type.

M257 Smoke Discharger
Electrically operated, the M257 fires a pattern of four smoke grenades in front of the Bradley as an emergency defence measure.

Engine compartment
The Cummins VTA-903T turbo-charged 8-cylinder diesel engine develops 500 hp at 2600 rpm. It is equipped with a Halon (inert gas) fire suppression system.

MECHANISED COMBAT

gas fire-suppression system but the gas they used, Halon, reacted with a fuel fire to produce toxic fumes. Not quite the intended result.

Improved warload

The M2 Bradley carries 900 25-mm cannon shells; 2,340 7.62-mm bullets; 4,200 5.56-mm bullets; and 10 TOW anti-tank missiles. The M3 carries substantially more of everything. Add 662 litres of fuel, and you can see why the US Congressional Report concluded that enemy shots would probably hit something vital and "catastrophic explosions are likely to occur with unacceptable frequency during combat".

Congress forced the Pentagon to conduct operational and live-fire tests on two modified versions of the Bradley. The tests did show the Bradley to be proof against small-arms fire, from 7.62-mm bullets up to the 14.5-mm ammunition used by Soviet heavy machine-guns. It was also proof against 155-mm shell fragments from overhead bursts, but it seems the Bradley is still vulnerable to shaped-charge infantry anti-tank weapons.

In 1985, the M2A1 Bradley went into production. This incorporated an

Above: The Bradley carries TOW anti-tank guided missiles, which have a maximum effective range of 3,750 m (12,300 ft). When on the move, the launcher is retracted and lies along the left-hand side of the turret.

Vehicle commander
The commander has periscopes for forward and side observation and a single-piece hatch opening backwards. He dismounts with the infantry squad.

TOW anti-tank guided missile launcher
Here, the twin TOW launcher has been elevated through 90° into the firing position. The sighting and controls for the 25-mm cannon and the TOW missiles are fully integrated to make the gunner's life easier. A thermal imaging sight provides ×4 and ×12 viewing for the gunner and the commander.

Gunner
Provided with a primary and a back-up sight, the gunner can choose between APDS-T (Armour Piercing Discarding Sabot-Tracer) and HEI-T (High Explosive Incendiary-Tracer) ammunition for the cannon as well as operating the TOW system and the co-axial machine gun.

Firing port
The infantrymen in the troop compartment are provided with firing ports, each with a periscope above. These enable them to fire M231 Personal Weapons from within the vehicle; accuracy is not high, but it allows the Bradley to deliver an incredible volume of suppressive fire.

Suspension
The Bradley has a torsion bar suspension system and six road wheels with rubber tyres. All wheels except the fourth and fifth are fitted with hydraulic shock absorbers.

MECHANISED COMBAT

The Bradley has one feature common to most Russian Armoured Fighting Vehicles; it can generate a smoke screen by injecting diesel fuel into the exhaust. Unlike the British MCV-80 the Bradley is fully amphibious, propelling itself through the water with its tracks.

updated NBC system, much improved stowage of spare 25-mm ammunition and TOW missiles, and upgrades to the fuel and fire-suppression systems. All M2s have received additional armour, pushing unit weight up from 22 to 27 tonnes (21.6 to 26.6 tons). This has made the machine's amphibious quality chancy, but has increased survivability. Bradleys are currently being tested with several varieties of passive and explosive reactive armour (ERA), intended to increase protection against enemy HEAT warheads.

Superior performance

The US Army has a requirement for 6,720 M2 Bradleys and its M3 cavalry variant, to equip its front-line mechanised units. The M2 saw extensive service in the Gulf, and, apart from some problems with the transmission system, worked reasonably well. In spite of all doubts the Bradley has proved to be a superior fighting vehicle, highly mobile and capable of taking on most potential opponents.

Infantry and Cavalry Fighting Vehicles

There are two versions of the Bradley: the M2, the basic Infantry Fighting Vehicle carrying a full squad of infantry, and the M3, a reconnaissance vehicle.

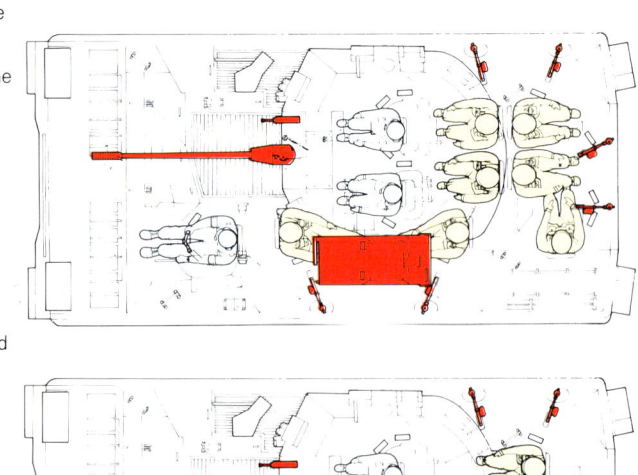

M2
The M2 carries seven infantrymen and the driver in its hull. The gunner sits in the left-hand side of the turret and the vehicle commander, who dismounts with the infantry, in the right-hand side. Six firing ports are provided so that the infantry can shoot from inside the vehicle.

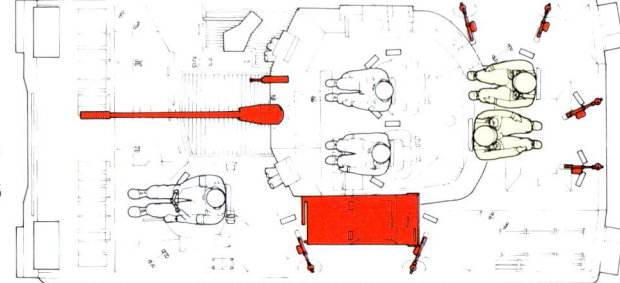

M3
Looking identical from the outside, the M3 carries a crew of only five but nearly twice as much ammunition, including 15 TOW anti-tank missiles, 1,200 25-mm cannon shells, and 4,500 7.62-mm bullets.

MECHANISED COMBAT

UP FRONT WITH THE VAB

The Renault VAB owes its development to the French insistence on relying as far as possible on the domestic market for the production of her military hardware. In the late 1960s the French army decided to issue its infantry units with both tracked and wheeled vehicles. Although tracked vehicles were obviously far stronger they were considerably more expensive to build, and required considerably more maintenance. It was thought wasteful to issue tracked vehicles to rear or support troops who were not tasked with meeting the enemy in head-on battle.

The French decided not to buy foreign equipment, although at the time there were several excellent armoured personnel carriers available within NATO. Development of the tracked AMX-10P MICV was already under way when, in 1970, requirements for a wheeled 'Forward Area' Armoured Vehicle were announced.

Army testing
Prototypes of 4×4 and 6×6 vehicles built by Panhard and Saviem/Renault were tested extensively by the army, and in May 1974 the Saviem/Renault Group 4×4 VAB was selected. The prototype was so good that pre-production vehicles were dispensed with, and the first production vehicle

It is the 14th of July, and the Bastille Day celebrations in Paris are in full swing. Infantry units mounted in VAB armoured personnel carriers head the French army's parade down the Champs Elysées. However, the very successful VAB is designed for much more difficult work than parading.

entered service in autumn 1976.

The French army required between 4,000 and 5,000 VABs. In June 1981 Renault announced that it had built 1,500 vehicles for the home and domestic markets and had orders for a further 5,000 assorted vehicles, which were being constructed at the rate of 50 per month.

The basic model in service with the French army is the 4×4 VAB VTT,

MECHANISED COMBAT

Inside the VAB

Both the 4×4 and 6×6 versions of the VAB have the same general structure. The hull is sufficiently armoured to protect the occupants from small arms fire, grenades and shell fragments. All are amphibious, and NBC systems, heating, de-frosting, and air-conditioning are all available, depending on the customer's requirements. This is a VTM 120 in a desert camouflage developed for models exported to the Middle East.

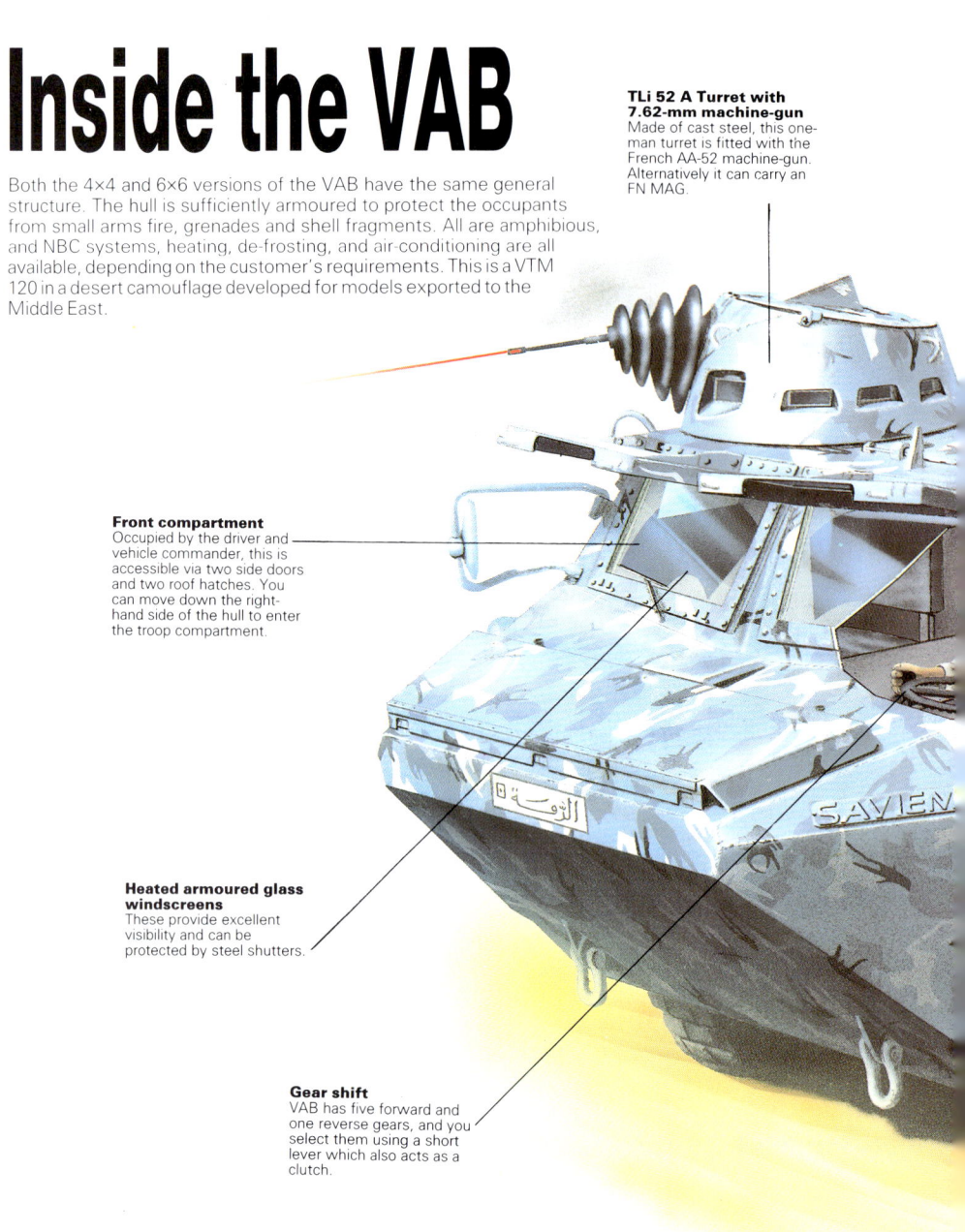

TLi 52 A Turret with 7.62-mm machine-gun
Made of cast steel, this one-man turret is fitted with the French AA-52 machine-gun. Alternatively it can carry an FN MAG.

Front compartment
Occupied by the driver and vehicle commander, this is accessible via two side doors and two roof hatches. You can move down the right-hand side of the hull to enter the troop compartment.

Heated armoured glass windscreens
These provide excellent visibility and can be protected by steel shutters.

Gear shift
VAB has five forward and one reverse gears, and you select them using a short lever which also acts as a clutch.

VABs are amphibious, propelled in the water by hydrojets at 7 km/h (4.4 mph). Before entering the water you must switch on the bilge pumps and raise the trim vane on the glacis plate. You steer the waterjets using a joystick on the dashboard.
which has a crew of two (commander/machine-gunner and driver) and passenger compartment for 10 infantrymen. The 5.98-m (19.6-ft) long all-welded steel hull provides protection from small arms fire and shell splinters, but would be of little use against concentrated heavy artillery bombardment or enemy tanks.

The driver sits to the front left of the vehicle with the commander/machine-gunner to his right. Both crewmen have small forward-opening doors for easy access. The front and door windows contain bullet-proof glass, and the front windows are heated. These can be covered by steel shutters if required, although this limits the driver's field of vision.

In its basic form the 4×4 VAB VTT is fitted with front-opening hatches above both crew positions, but all French models have had a small Creusot-Loire rotating gun mount fitted above the commander's seat. Early models were armed with a 7.62-mm machine-gun capable of 360-degree traverse and an elevation of −15 to +45 degrees conventionally and from −20 to +80 degrees in the anti-aircraft mode, but current models are now fitted with a 12.7-mm Browning M2HB heavy machine-gun.

A passageway on the right side of the hull connects the crew compartment with the passengers seated in the rear. The infantrymen enter and leave the vehicle via two outward-opening rear doors, each of which has a window that can be covered by an armoured shutter.

Three firing ports are provided on each side of the hull. Although these can be secured open externally to

A VAB undergoes a fire test. Quiet and relatively unobtrusive, it makes a good IS (Internal Security) vehicle. Note that the armoured shutters have been lowered to protect the windscreen.

MECHANISED COMBAT

120-mm mortar
Demonstrated at Satory in 1986, the VPM 120 model of the VAB carries a Thomson-Brandt MO 120 LT mortar. Crew consists of the driver, the commander (who operates the machine-gun turret as well as commanding the vehicle), and four mortar crew in the crew compartment.

Troop compartment
In the APC version the seats are along the side and fold up to allow cargo to be stored instead. Because there is no central pillar the full capacity of the troop compartment can be used and a fork-lift truck may be employed to load the cargo.

Gunport openings
These are protected by armoured glass which can be locked open so that you can fire rifles from within the troop compartment.

Michelin run-flat radials
The tyre pressure can be varied according to the sort of terrain the VAB is travelling across.

Wheels
The wheels are independently suspended by torsion bars and telescopic shock absorbers.

Engine compartment
This houses a Renault VI MIDS 6-cylinder in-line water-cooled turbo-charged diesel which develops 235 hp.

enable the passengers to fire their small arms on the move, the soldiers are seated on benches running along both sides of the hull (as opposed to back-to-back, facing outwards, as in the case of the Russian BMP), so accurate marksmanship is virtually impossible. A circular hatch, to which a variety of armament installations can be fitted, is situated above the forward part of the upper hull, with two smaller hatches opening forward, to its rear.

The troop compartment is spacious, to reduce troop discomfort as far as possible: particularly important when you remember that reinforcements, many of whom will spend a considerable time travelling, will have to be fit and fresh when delivered to the battlefield. If necessary the seats can be folded to enable up to 2,000 kg (4,409 lb) of cargo to be carried.

The engine, transmission and 300-l (66-gal) fuel tank are situated together behind the driver and offset to the left, enabling the crew access to the crew compartment. French models are fitted with a MAN D 2356 HM 72 six-cylinder inline water-cooled diesel developing 235 hp at 2,200 rpm, but since 1984 this has been replaced in export models with the Renault VI MIDS 06.20.45 six-cylinder inline water-cooled turbo-charged diesel engine, developing 230 hp at 2,200 rpm.

Power is transmitted to the wheels by a hydraulic torque converter and transmission with five forward and one reverse gears, and a small pneumatically-operated lever operates both the gears and clutch. The wheels are independently suspended by torsion bars and hydraulic shock absorbers. All wheels are run-flat with pres-

The French army uses the VCAC anti-tank version, which is a 4×4 model of the VAB fitted with the Euromissile Mephisto system. It has four ready-to-fire HOT missiles and the launcher can be retracted when not in use so the vehicle looks like an ordinary APC.

179

MECHANISED COMBAT

The interior of the VTM 81, which has an 81-mm mortar fitted on a turntable with 360 degree traverse and has a range of 5 km (3.12 miles). The VTM 120 is similar, carrying a Thomson-Brandt 120-mm mortar with a range of 7 km (4.4 miles) and much more powerful ammunition.

sure controlled internally, and the front pair (front four in the case of the 6×6) are hydraulically assisted.

The versatile VAB is capable of a top speed of 92 km/h (57 mph), has an excellent road range of 1,000 km (620 miles), can climb gradients of 60 per cent and will be an ideal way of getting reinforcements and supplies to France's forward divisions in Germany.

Amphibious properties

Apart from a few early French army models, VABs are fully amphibious and capable of a maximum water speed of 7 km/h (4.4 mph). Propulsion comes from two Dowty waterjets fitted to the rear of the hull. Both jets are fitted with a deflector for steering and reverse thrust and are hydraul-

Below: Capable of over 90 km/h (56.25 mph) and with ranges of 1,100-1,300 km (683–800 miles), the VAB is highly mobile. Independently suspended wheels with large clearance helps VABs cope with 60 per cent gradients and 30 per cent slopes.

ically controlled by a small joystick mounted on the dashboard.

Before entering the water, bilge pumps are switched on, and the trim vane, which at all other times is folded back on the glacis plate, is erected on the front of the boat-shaped hull.

With their usual eye on the Third World export market, the French have designed the basic VAB as little more than a relatively cheap shell to which a series of extras can be added when finances permit. Optional equipment includes an NBC system (fitted to all French army models), infra-red or passive night vision equipment, an air-conditioning system, gas dispensers and grenade throwers. A front-mounted winch with a capacity of 7,000 kg (15,432 lb), doubled with the aid of a pulley, and 60 m (196.8 ft) of cable can be fitted to give the vehicle a basic engineering capability.

The spacious VAB can easily be adapted. Although never intended as a front-line combat vehicle, a VCI

The VAB ambulance is designed to provide rapid aid to wounded men directly on the battlefield. It can carry either four stretcher cases, 10 sitting wounded or two stretcher cases and five sitting wounded. It has a ventilation system for greater comfort.

Infantry Combat Vehicle (Véhicule de Combat de L'Infantrie) has been developed for export and for use by the French air force for air field defence. Manned by a crew of three (commander, driver and gunner) and capable of carrying eight fully-equipped infantrymen, the VCI is armed with a centrally-mounted 20-mm cannon and 7.62-mm co-axial machine-gun.

Anti-tank variant

Approximately 60 VAB HOT anti-tank vehicles carrying the Euromissile Mephisto system with four ready-to-launch HOT missiles are in French service, and a similar system fitted with a UTM turret has been exported to Qatar. Capable of destroying any APC and all but the latest generation of tanks, the HOT missile with its maximum range of 4,000 m (13,120 ft) remains a formidable weapon.

Two mortar vehicles have appeared: the VTM (Véhicule Tracteur de Mortier), which tows a Brandt 120-mm mortar, and a prototype which fires an 81-mm mortar through a two-part opening in the roof. Command, artillery control and surveillance radar variants complete the series so far.

The VAB is without doubt one of the greatest success stories of French military design. Construction will continue for many years to come, and new variants will no doubt continue to come onto the export market. The VAB is cheap and easy to produce and simple to maintain and drive, and can fulfil most non-combatant roles.

MECHANISED COMBAT

MOWAG

Armoured Personnel Carrier

Above: The MOWAG Piranha 4x4 is seen here with an externally-mounted Browning M2 12.7-mm machine-gun. Other armament options include twin 7.62-mm machine-guns.

Since the end of World War II the MOWAG company has manufactured a wide range of tracked and wheeled armoured fighting vehicles aimed mainly at the export market, and has also built prototypes of armoured vehicles for foreign governments. For example, MOWAG built some of the prototypes of the West German Marder mechanised infantry combat vehicle. In the 1950s a 4×4 series of armoured vehicles was designed and built under the company designation MOWAG MR 8, and this was subsequently adopted by the West German border police in two configurations, the SW1 and the SW2. The first batch of 20 or so vehicles was supplied direct by MOWAG, but main production was undertaken in West Germany by Henschel and Büssing. Total production in West Germany amounted to about 600 vehicles.

Above: The Piranha series of vehicles is fully amphibious, powered in the water by twin propellers at the rear of the hull. The 8x8 type seen here is now in service with the US Marine Corps.

The SW1 (*geschützter Sonderwagen Kfz 91*) is the armoured personnel carrier model and accommodates five men plus the commander and driver, while the SW2 has a slightly different hull top and is fitted with a one-man turret armed with a 20-mm Hispano-Suiza cannon plus four smoke dischargers mounted on each side of the turret to fire forwards.

The same basic hull is used for both the SW1 and SW2, with slight differences to the roof. In the SW1 the commander and driver are seated at the front of the hull with a windscreen in front of each man; these windscreens

181

MECHANISED COMBAT

can be quickly covered by armoured shutters with integral vision blocks. The driver also has a roof hatch above his position for driving in the head-out position. The troop compartment is at the rear of the vehicle, with the engine compartment to its left. In each side of the hull is a two-part door that opens left and right; each door has a vision block and a firing port. Over the top of the troop compartment are two roof hatches and an unusual cupola. The latter is fixed but split down the middle so that it can be opened vertically if required; in each half are three fixed vision blocks. When the cupola is in the normal position complete visibility is possible through 360 degrees.

Unlike more recent MOWAG wheeled armoured vehicles, the MR 8 series of vehicles have no amphibious capability and are not fitted with an NBC system or any type of night vision equipment, although both of the latter can be fitted if so required by the user.

MOWAG continued to develop the MR 8 series for other export markets, and these variants included the MR 8-09 sporting a one-man turret armed with a 20-mm cannon, the MR 8-23 that had a two-man turret armed with a 90-mm gun and a 7.62-mm co-axial machine-gun, and the MR 9-32 fitted with a 120-mm mortar at the rear of the hull. The last version had an open-top hull, and before the mortar could be fired it had to be lowered to the ground.

Above: The Piranha's suspension is independent, with coil springs on the forward pair of axles and torsion bars on the rear. All wheel stations are fitted with hydraulic shock absorbers.

MOWAG Roland

The MOWAG Roland 4×4 is the smallest vehicle currently produced by the MOWAG company and is used mainly in the internal security role. The first prototype was completed in 1963, the first production vehicles being completed the following year. Known operators of the Roland include Argentina, Bolivia, Chile, Greece, Iraq, Mexico and Peru. The hull of Roland is of all-welded steel armour construction that provides the crew with complete protection from 7.62-mm small arms fire. The driver is at the front, the crew compartment in the centre and the engine at the rear on the left side; there is also an aisle in the right side of the hull that leads to a door in the hull rear. The driver has a roof hatch, and there is a single door in each side of the hull. In each of the three doors is a firing port (with a vision block above) which allows three of the embarked infantrymen to fire their rifles or sub-machine guns from within the vehicle in safety.

In the centre of the roof is installed the main armament; this is normally a simple cupola with an externally mounted 12.7-mm or 7.62-mm machine-gun. One of the alternative weapon stations is a turret, on top of which is a remotely-controlled 7.62-mm machine-gun fired from within the turret.

The petrol engine is coupled to a manual gear-box with four forward

Above: The Roland internal security vehicle is widely used in Africa and South America. This version has a remote 7.62-mm machine-gun turret.

Below: The Roland can serve as a command vehicle, ambulance and cargo/ammunition vehicle. Early models had a manual transmission; later ones were automatic.

MECHANISED COMBAT

and one reverse gear and two-speed transfer case. More recent production Rolands are offered with an automatic gear-box to reduce driver fatigue.

When used in the internal security role, the Roland is normally fitted with an obstacle-clearing blade at the front of the hull, a public address system, wire mesh protection for the headlamps and sometimes the vision blocks as well, a siren and flashing lights. Another option is MOWAG bulletproof cross-country wheels. These consist of metal discs on each side of the tyre, the outside ones having ribs which assist the vehicle when crossing through mud.

Grenadier APC

In the late 1960s the company designed and built another 4×4 armoured personnel carrier called the MOWAG Grenadier, which can carry a total of nine men including the commander and driver. This model was sold to a number of countries but is no longer offered, having been replaced by the Piranha range of 4×4, 6×6 and 8×8 armoured vehicles. Typical armament installations for the Grenadier included a one-man turret armed with a 20-mm Hispano-Suiza cannon and a turret with twin 80-mm rocket launchers. The vehicle is fully amphibious, being propelled in the water by a propeller under the rear of the hull. Waterborne steering is accomplished by turning the steering wheel in the normal manner to move two parallel rudders mounted to the immediate rear of the propeller.

The MOWAG Piranha range of 4×4, 6×6 and 8×8 armoured personnel carriers was designed by MOWAG in the late 1960s, and the first prototype was completed in Switzerland in 1972, with first production vehicles following four years later. As with all recent MOWAG vehicles, the Piranha family was a private venture and developed without government support.

In 1977 Canada decided to adopt the 6×6 version and production was undertaken in Canada by the Diesel Division of General Motors Canada,

The 4x4 version of the Piranha has been sold to several West African countries, including Liberia where they have taken part in the civil war. It is unlikely that any are still operational.

A clear view of a MOWAG Piranha with 20-mm cannon and co-axial 7.62-mm machine-gun. Note the defensive smoke grenade launchers on the turret side and the propeller above the rear wheel.

MECHANISED COMBAT

491 being built for the Canadian Armed Forces between 1979 and 1982. Canada uses three versions of the 6×6 Piranha: the 76-mm Cougar Gun Wheeled Fire Support Vehicle, which has the same two-man turret as the British Combat Vehicle Reconnaissance (Tracked) Scorpion; the Grizzly Wheeled Armoured Personnel Carrier, which has a one-man turret armed with a 12.7-mm and a 7.62-mm machine-gun and has a three-man crew consisting of commander, gunner and driver plus six fully equipped troops; and the Husky Wheeled Maintenance and Recovery Vehicle, which supports the other vehicles in the field.

In addition to being used by Canada, the Piranha range of vehicles is used also by Chile (licence production), Ghana, Nigeria and Sierra Leone, and in 1983 the 6×6 model was evaluated by the Swiss army as an anti-tank vehicle fitted with the Hughes TOW anti-tank system. After evaluating a number of different vehicles, both tracked and wheeled, the USA selected the 8×8 version of the Piranha to meet its requirement for a Light Armored Vehicle (LAV) and the first of these was completed for the US Marine Corps in late 1983. These have a two-man power-operated turret armed with a Hughes Helicopters 25-mm cannon (as fitted to the Bradley) and a co-axial 7.62-mm machine-gun. Variants required by the US Marines include a logistics support vehicle, a command vehicle, a repair vehicle, a mortar carrier and an anti-tank model.

Steel construction

The hull of the Piranha is of all-welded steel construction, which provides protection from small arms fire. On the six-wheeled version the driver is at the front on the left with the commander to his rear and the engine to the right. The troop compartment is at the rear of the hull, and entry to this is gained via two doors in the hull rear. Armament depends on the role, but can range from a single-man turret up to a two-man power-operated turret armed with a 90-mm Cockerill gun. If a heavy weapon such as this is fitted, however, the commander is normally in the turret and a reduced number of troops is carried.

All members of the Piranha family are fully amphibious, being propelled in the water by two propellers at the rear of the hull. Optional equipment includes night vision equipment, an NBC system and an air-conditioning system.

Below: With a power-to-weight ratio of over 27 hp per ton, the Piranha has plenty of power in reserve to tackle cross-country operations.

MECHANISED COMBAT

FORWARD WITH THE SPARTAN

The highly successful Scorpion CVR(T) is now just one member of a family of excellent light armoured vehicles. Serving in a wide variety of roles, they offer the British Army a tracked weapons platform with excellent battlefield mobility. All the different versions are powered by the same military model of the Jaguar OHC 4.2-l (0.9-gal) petrol engine, derated from 240 to 195 horsepower. A GM diesel can be fitted to give extra range and reduced fire hazard.

A TN-15X cross-drive transmission, developed from that originally introduced into the Chieftain tank, provides seven speeds in each direction. For engine cooling, a single mixed-flow fan draws in air through the radiator over the gear-box, over the engine and out through the louvres.

The first Scorpion variant to enter service was the Striker anti-tank guided weapon vehicle introduced in June 1975 as a replacement for the FV 438. First deployed with the British Army of the Rhine (BAOR) a year later, Striker was issued originally to the Royal Artillery but has now been issued to the Royal Armoured Corps, eight vehicles to each regiment.

The driver, who sits forward left next to the engine, is immediately in front of the commander, who has the missile controller to his right. The commander has a low cupola, with eight periscopes offering all-round view with visibility enhanced by a monocular sight offering ×1 and ×10 magnification.

Armament

A 7.62-mm machine-gun, which can be aimed and fired from within, is situated to the right of the cupola and gives some capacity for self-defence. The missile controller, who must maintain line of sight with his target when engaging, has a split-view monocular sight with ×1 and ×10 magnification capable of traverse through 55

The FV 103 Spartan is a small armoured personnel carrier based on the chassis of the CVR(T) Scorpion/Scimitar. Too small for an infantry section, it is used to carry Royal Artillery surface-to-air missile teams and also serves with the RAF Regiment.

degrees left and right.

The teeth of the vehicle, the five British Aerospace Swingfire Anti-Tank Guided Missile launchers, are contained in a box on the upper body to the rear of the cupola. A refill of five missiles is carried inside the vehicle, but can only be loaded manually by a crewmember: a very real problem if under fire or under NBC conditions.

The launcher box, which is pivoted to the rear and raised to an angle of approximately 35 degrees before firing, can be controlled by the firer either from his seat inside the vehicle or from a remote position up to 100 metres away, enabling the Striker to be hidden behind cover when firing.

MECHANISED COMBAT

Above: The CVR(T) chassis makes an excellent tank destroyer when fitted with Swingfire anti-tank guided missiles. The Striker's missiles can be guided from outside the vehicle, giving a useful indirect fire capability.

Below: Sultan deployed in a suitably wooded area. The vehicle and its tent should be camouflaged in the field as any command post will be a primary target. A radio antenna can be erected at the front of the vehicle when it is in a static position.

The Euromissile consortium which makes MILAN sold the British Army 75 MCT compact turrets in 1985. Fitted to the CVR(T) chassis, they provided an alternative anti-tank system to Striker.

Although this has its obvious advantages, the firer remains vulnerable to suppressive fire and would be unwise to remain in one position very long.

Swingfire missile

The Swingfire missile, which weighs 28 kg (61.7 lb) and has a HEAT warhead, has a minimum range of 150-300 m (492–984 ft), depending on the situation of the firer at the time. Its maximum range is stated as 4,000 m (13,120 ft), although the chance of hitting anything but a large static target at that sort of range are fairly limited. In live firing training, now something of a rarity, the crews tend to concentrate on targets less than half that distance.

Recent innovations in tank ar-

Inside the Spartan

Spartan is a useful armoured personnel carrier for those units which operate in smaller groups than the eight-man infantry section. In addition to army use, it now serves with the RAF Regiment in the airfield defence role.

Driver's periscope

Smoke grenade launchers

Aluminium armour
Proved to be surprisingly effective against shell splinters, this is vulnerable to heavy machine-guns, and even 7.62-mm from a GPMG can penetrate from certain angles.

Driver's hatch
The driver is out of sight, driving closed down. He relies on his single wide-angle periscope to see where he is going. This can be replaced with a passive night sight.

Tracks
The rubber-bushed steel tracks have expectancy of mo 5,000 km (3,107 r

MECHANISED COMBAT

moured protection, notably Russian reactive armour, reduce the value of most anti-tank guided weapon systems. But Swingfire, and with it Striker, remain an integral part of British anti-armour thinking, and are likely to do so for many years to come.

Spartan APC

The FV 103 Spartan Armoured Personnel Carrier was the least successful of the family when it first appeared, because it was simply too small to act in the intended role of infantry section troop carrier. Entering service in BAOR in 1978, it has since been deployed in a number of support roles, including carrying Royal Artillery Blowpipe/Javelin SAM teams, transporting Royal Engineer assault teams, and resupplying Striker. More recently it has been adopted by the RAF Regiment and now plays a key part in airfield defence.

Spartan, which can carry four fully laden passengers in addition to the driver, commander/gunner and section commander/radio operator, has a similar hull configuration to Striker.

Entry to the rear personnel section is through a single door hinging to the right. Vision blocks are provided to help alleviate passenger disorientation, but firing blocks are not fitted, although four small hatches are provided above the troop compartment.

Attempts have been made to fit Spartan with a number of weapon systems, including twin 20-mm anti-aircraft cannon fitted in the French ESD turret, a Hughes TOW (Tube Optical Wired) launcher and a Euromissile HOT ATGW system.

The most practical of these attempts bore fruit in 1980 when Spartan was successfully fitted with the Euromissile MILAN ATGW system in a specially designed MCT compact turret. The system allows the missiles to be launched by the gunner from within the vehicle. Operation and guidance remain identical to the original MILAN infantry system and a night sight allows the crew to engage enemy armour during the hours of darkness. Each Spartan carries two MILANs ready to fire, with a further 11 inside the manual reloading, seven on the left and four on the right.

To date a limited number of systems

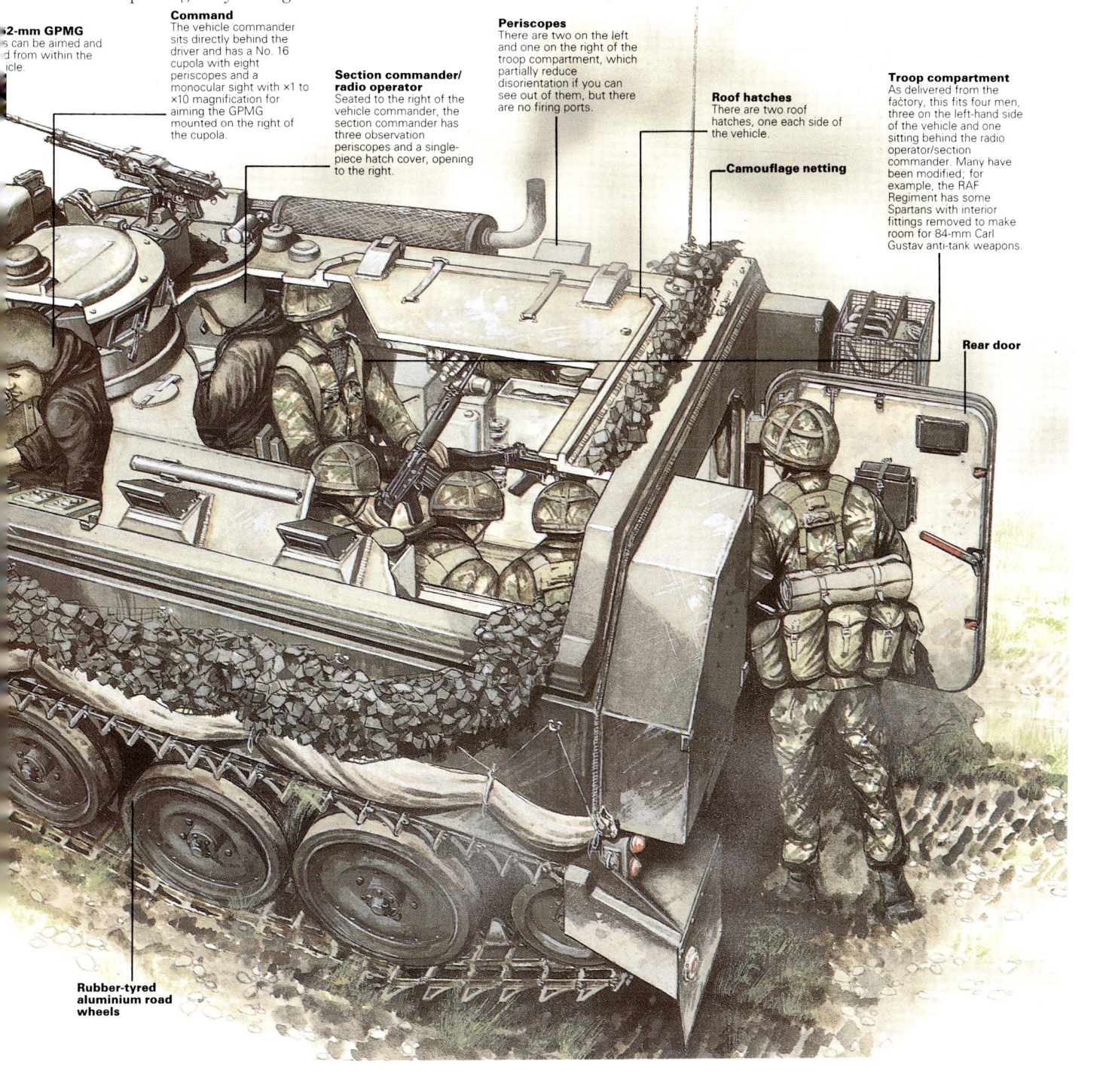

2-mm GPMG
s can be aimed and d from within the icle.

Command
The vehicle commander sits directly behind the driver and has a No. 16 cupola with eight periscopes and a monocular sight with ×1 to ×10 magnification for aiming the GPMG mounted on the right of the cupola.

Section commander/radio operator
Seated to the right of the vehicle commander, the section commander has three observation periscopes and a single-piece hatch cover, opening to the right.

Periscopes
There are two on the left and one on the right of the troop compartment, which partially reduce disorientation if you can see out of them, but there are no firing ports.

Roof hatches
There are two roof hatches, one each side of the vehicle.

Camouflage netting

Troop compartment
As delivered from the factory, this fits four men, three on the left-hand side of the vehicle and one sitting behind the radio operator/section commander. Many have been modified; for example, the RAF Regiment has some Spartans with interior fittings removed to make room for 84-mm Carl Gustav anti-tank weapons.

Rear door

Rubber-tyred aluminium road wheels

MECHANISED COMBAT

Right: The Sultan armoured command vehicle has a hull similar to that of the Samaritan. The room available for the command post is extended by erecting this tent at the rear of the vehicle. This Sultan is in service with the Belgian army.

have entered service with British infantry battalions, but it is unlikely that the concept, with its relatively short range, will ever be fully accepted.

Stormer, in essence a stretched version of the Spartan, was introduced to carry a full section of eight infantrymen, but with the advent of Warrior it is unlikely to be adopted by the British Army.

Samaritan

The FV 104 Samaritan armoured ambulance entered service in 1978 and has been a great success. Manned by a crew of two, the driver and commander/medical orderly, Samaritan is completely unarmed, relying upon its large red cross for protection.

Rather idealistically, the red crosses are fitted with cloth covers which may be lowered to cover the emblems if the vehicle is used in a combatant role. The wounded, who may be transported as four stretcher cases, five walking wounded or two stretcher and three sitting cases, enter the vehicle via a single large door at the rear.

Samaritans are attached to every mechanised infantry battalion and armoured regiment and can traverse the roughest of conditions. With the assistance of a passive night sight they can operate as effectively in darkness as daylight.

Sultan

The first production FV 105 Sultans entered service in April 1977 and quickly replaced the elderly Saracen as the basic command vehicle; they now operate throughout the Army in roles as diverse as artillery command posts and divisional headquarters support vehicles. Considerably higher than the original Scorpion, the Sultan can be quickly fitted with a canvas penthouse attached to the rear of the chassis for additional room. The vehicle is usually manned by a crew of five or six, the driver, commander, radio operator and two or three specialists.

Normally equipped with at least two radios, one at the front and one in the rear working area, Sultan is readily identifiable by a radio antenna often erected at the front of the vehicle when stationary. A single 7.62-mm machine-gun is pintle-mounted on the roof for minimal anti-aircraft protection. Attempts are under way to fit specialist Sultans with advanced electronic warfare systems although, in line with all aspects of electronic warfare, precise details remain shrouded in secrecy.

Samson

The Samson ARV entered service in 1978 and has since operated as a highly efficient recovery vehicle for all tracked armoured vehicles other than tanks (which it is too small to move). A heavy-duty winch fitted inside the hull controls a 229-m (751-ft) wire rope, has a variable speed of up to 122 m (400 ft) per minute and is driven from the main engine.

When in use for recovery, the winch is supported by two manually operated spades lowered at the rear. An A-frame, fitted at the rear of the hull, is capable of lifting a tank engine if need be, enabling complex field repairs to

Above: This Belgian army Samson armoured recovery vehicle has its spades lowered and A-frame erected. The spades stabilise the vehicle when using its winch, which has a maximum pull of 12,000 kg (26,455 lb).

be made. This is a vital service taking into account the limited number of armoured vehicles, especially tanks, available to the British Army.

The future

The CVR(T) has proved a major success, marred only by the limited number of vehicles in service. Unfortunately the emphasis in armoured construction has now passed to the Warrior, with the result that the lack of these excellent vehicles is likely to remain a problem.

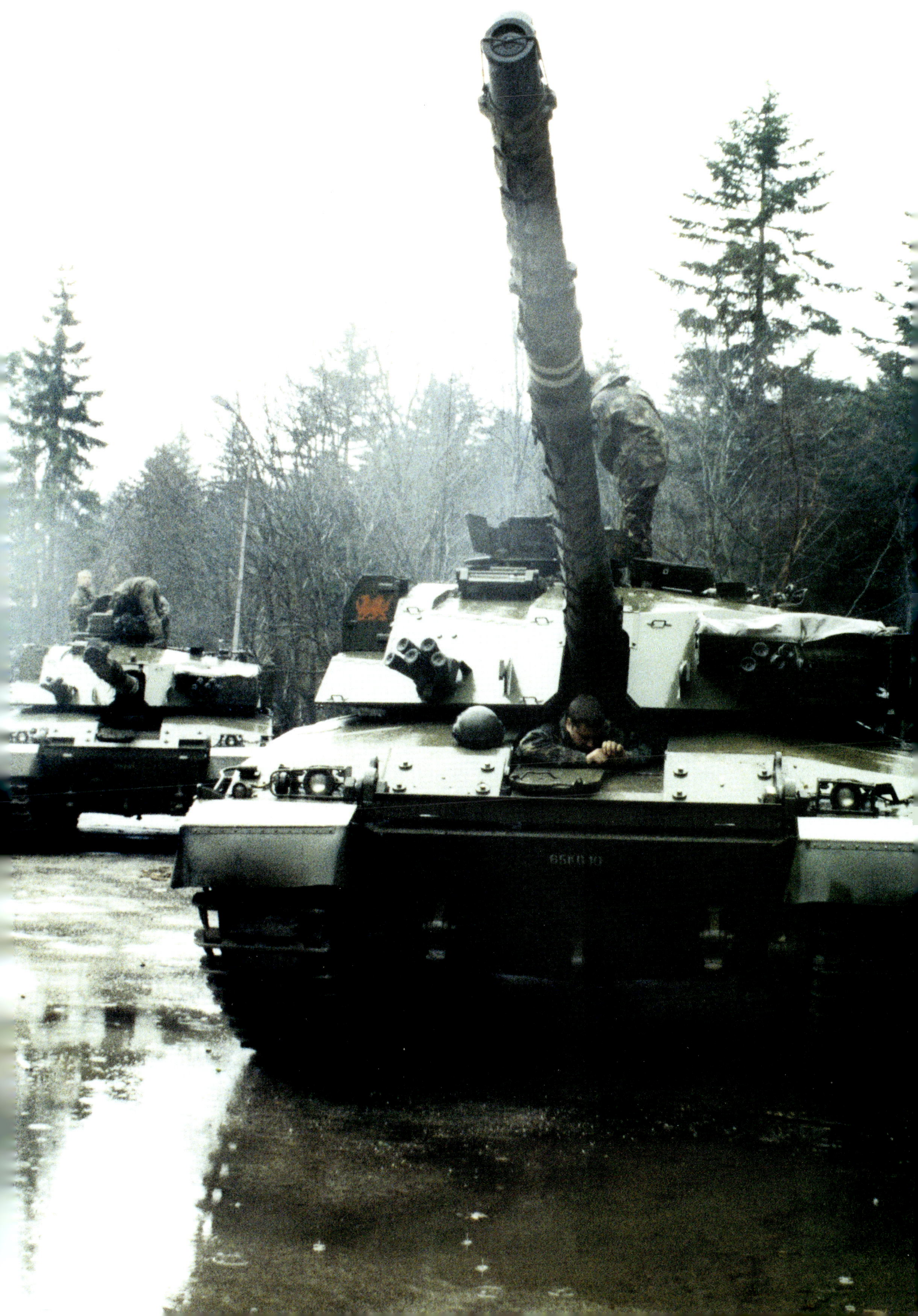

Index

Page numbers in *italics* refer to illustrations

ACAVs *see* Armoured Cavalry Assault Vehicles	
advance to contact	17-20
airborne forces, BMD	53-6
aircraft	
A-4 Skyhawk	*35*
Buccaneer	*98-9*
covering forces	*123*
Mirage	*97*
see also helicopters	
all-terrain vehicles	
Dragoon	133-6
Ratel	77-80
ambushing	49-52, 60, 97-100
ammunition	
Armour-Piercing Discarding Sabot (APDS) rounds	
	16, 38
Armour-Piercing, Fin Stabilized, Discarding Sabot	
(APFSDS) rounds	38, *133*
high-explosive anti-tank (HEAT) rounds	
	105, 106, 107, 154
high-explosive squash head (HESH) anti-tank rounds	
	9-10, *11,* 15-16, 37-8, 150-2
sabot fin-stabilised anti-tank rounds (Fin)	9-11
amphibious assault vehicles	
AAV7	117-20
AMX-10P	158, 160
BMD	53-6
BMP	*153,* 155-6
Commando	171
Dragoon	134
FV-432 APC	110-11
Luchs	*102,* 104
LVTP series	117-18
M113	165
MOWAG	183-4
PT-76	*131*
VAB	*178,* 180
AMVs *see* anti-mine vehicles	
anti-aircraft weapons, ZSU-23-4	137-40
anti-mine vehicles (AMVs)	161-2
anti-tank ditches	25, *26*
anti-tank guided weapons (ATGWs)	33-4, 122, 126
APCs *see* armoured personnel carriers	
APDS rounds *see* ammunition	
APFSDS rounds *see* ammunition	
Argentina, Tanque Argentino Mediano	96
armaments	9-10, 58-9, 115, 121-2
AAV7	118-20
AGS-17 grenade launcher (*Plamya*)	56
AML armoured cars	106, *107,* 108
AMX-10P	158-60
ASU-57 self-propelled anti-tank gun	*54*
BMD	53-4
BMP	154-6, 173-4
Bradley	173-5
Buffel	162-3
Carl Gustav recoilless rifle	*58-9, 66*
Cascavel	85-8
Challenger	37-9
Chieftain	13-15
Commando	171-2
D-30 1.22-mm towed field howitzer	*54*
Dragon system	*65*
Dragoon	*133,* 134-6
Eland	*105,* 106-7
FAMAS 5.56mm rifles	*159*
grenades	52, 100
L7 gun	5-7
L11AS rifled gun	14
Light Anti-Tank Weapons (LAWs)	*50-1,* 58-60, *68,*
	127
Luchs	*102-3,* 104
M1 Abrams	30-1
M1A1	30
M46	*123*
M60	5-8, *34*
M73 machine-gun	8
M77	46
M86 machine-gun	8
M109 howitzers	145-8
M113	165, 167
M185 155mm howitzer	*145,* 147
M992 FAASV	*148*
McDonnell Douglas 25-mm Chain Gun	173
Marder	94-6
Medium Anti-Tank Weapons (MAWs)	59-60
Merkava	22-4
MOWAG	182, 184
Multiple Launch Rocket Systems (MLRS)	45-8, *124*
Plamya see AGS-17 grenade launcher	
RARDENS cannon	122, 125-7, *131,* 132
Ratel	78-80
Rheinmetall Mk20 Rh-202 cannon	95-6
RPG-7 anti-tank	174
Saladins	149-52
Saxon AT105	142-4
Scorpions	*130*
Sheridan M551	*130*
Spartan	187
Striker	185-6
T-54/T-55 series	61-4
VAB	178, 180
Warriors	125-7
Wombat 120mm anti-tank weapon	*124*
ZSU-23-4	137-40
armour	57
APCs abilities	114-15
appliqué type	168
Chobham	37, 39-40
armoured ambulances	188
armoured cars	
AML	105-8
BRDM-2	132
Cascavel	85-8
Eland	*100,* 105-8
FV601(D)	152
Luchs	101-4
Saladins	149-52
Armoured Cavalry Assault Vehicles (ACAVs)	173
armoured command vehicles	
Ratel	78
Sultan	188
armoured personnel carriers (APCs)	
AMX-10P	157-60
Boyevaya Mashina Desantniye (BMD)	53-6
Boyevaya Mashina Pekhoty (BMP)	
	49, 52, 57, *103,* 104, 153-6, 173-4
Bradley	168, 173-6
Casspirs	*98,* 100

combat skills	81-4
Commando	171
FV 432	74, 109-12, 128
M113	116, 133, 165-9
M577	167
movement strategies	113-16
MOWAG	181-4
Saxon AT105	141-4
Spartan	76, 187-8
Striker	66
VAB	177-80
see also mechanised combat vehicles	
armoured reconnaissance	129-32
armoured recovery vehicles (ARVs), Samson	188
Armoured Vehicle Royal Engineers (AVREs)	35
Armoured Vehicle-Launched Bridges (AVLBs), M60	8
artillery	45-8, 54, 67, 68, 124
ARVs *see* armoured recovery vehicles	
ATGWs *see* anti-tank guided weapons	
Australia, M113	168
AVLBs *see* Armoured Vehicle-Launched Bridges	
AVREs *see* Armoured Vehicle Royal Engineers	
blindfire	70-1
BMD *see* tanks	
BMP *see* Boyevaya Mashina Pekhoty	
bomblets, M-77	46-8
booby traps	99-100
Brazil, Cascavel	85-8
call signs	43-4
camouflaging	89-92
6CBAC *see* Combat Brigade Air Cavalry	
CEVs *see* Combat Engineer Vehicles	
Combat Brigade Air Cavalry (6CBAC)	73
Combat Engineer Vehicles (CEVs), M728	8
Combat Vehicle Reconnaissance (CVR(T))	121, 129-31
Scimitar	76, 121-2, 129, 131
Scorpion	129, 130, 185
Spartan	121-2
Striker	121-2, 123, 185
command posts, construction	27-8
Commando	169-72
communications	42-4, 83
counter-mobility	25-8, 36
counter-penetration	34-5, 73-6
covering forces	121-4
CVR(T) *see* Combat Vehicle Reconnaissance	
dead ground	59, 66
Dragoon	133-6
Egypt	21, 64
electronic warfare, Dragoon	136
Falkland Islands	117-18, 131
field defences	26-8
fire support	56, 78, 130-1
flares	52
formations, vehicles	17-18, 82-3
France	105-8, 177-80
Germany	10, 93-6, 152, 181-4
Great Britain	
Air Mobile Brigade	76
Challenger	9, 11, 16, 17, 35, 37-40, 125, 129

Chieftain	9-10, 13-16, 110
FV-432	109-12, 128
Marrior MCVs	112, 125-8, 132
Rapier	69-72
Saladins	149-52
Saxon AT105	141-4
Spartan	185-8
half-loading	9-10
helicopters	36, 74, 76
Apache	35, 67
Gazelle	129
Lynx	68, 75
HIP *see* Howitzer Improvement Program	
Howitzer Improvement Program (HIP)	148
howitzers, US M109	145-8
hull-down positions	115, 126
individual weapon sight (IWS)	59
Indonesia	152, 160
infantry	41-4, 65-7
airborne support	56
ambushing	49-52
APCs	81-4, 173-4
counter-penetration	74-5
infra-red signature	90
Internal Security (IS) vehicles	
British Saxon AT105	141-4
MOWAG	182-3
Iran, Challenger	37
IS *see* Internal Security	
Israel	21-4, 63-4, 168
IWS *see* individual weapon sight	
lasing	12
Libya, ZSU-23-4	140
Mechanised Infantry Combat Vehicles (MICVs)	
Marder	93-6
Warrior	35
Warriors	112, 125-8, 132
MICVs *see* Mechanised Infantry Combat Vehicles	
mine-protected fighting vehicles (MPFVs)	
Buffel	161-4
see also Ratel	
mine-protected vehicles (MPVs)	161
mines	100
anti-personnel	28
anti-tank	26, 50-2, 67-8, 143
see also anti-mine vehicles	
missiles	
Anti-Tank Guided Missile	84
AT-4 'Spigot'	154
HOT	66-7, 180
MILAN	27, 28, 66-8, 76, 187
'Sagger' anti-tank	21, 53
Swingfire	66-7, 186-7
TOW anti-tank	35, 36, 66-8, 73, 75, 114
Bradley	174, 175
Commando	172
M113	167
see also surface-to-air missiles	
mobile defence, tanks	33-6
mobile operations	97-100
mobile optical fire unit (MOFU)	71-2
MOFU *see* mobile optical fire unit	
mortar carriers	136, 172
mortars, M113	166-7

movement strategies, APCs	113-16
MOWAG	181-4
MPFVs *see* mine-protected fighting vehicles	
MPVs *see* mine-protected vehicles	
multi-mission vehicle, Commando	169-72
NATO	73, 93
night combat	22, 24, 32, *116*
Nuclear/Biological/Chemical warfare	110, 156
Pakistan, T54/T55 series	64
radars	69-71, 96, 137-40
radios	42-4, *91*
Ratel	77-80
reference points	44
road blocks	25-6, *28*, *51*
SADF *see* South African Defence Force	
Sappers, counter mobility	26-8
Schulzenpanzer Roland	96
scouting	80
shell strike method	44
signals	*115*, *116*
smoke generators, US M1059	167-8
South Africa	77-80, *100*, 105-8, 161-4
South African Defence Force (SADF)	99, *100*
Soviet Union	
BMD	53-6
BMP	*49*, *52*, *57*, *103*, 104, 153-6, 173-4
T-54 series	61-2
T-55 series	61-4
T-62	62, *65*
ZSU-23-4	137-40
special forces, camouflaging	89-92
street fights, APCs	115
submunitions, MLRS	46
supplies	79-80, 83-4
surface-to-air missiles	
Laserfire	72
Rapier	69-72
SA-7 'Grail'	54-6
48 SA-9 'Gaskin'	54
TAM *see* tanks	
tank gun barrel method	44
tank hunting	57-60
tanks	
Challenger	9, *11*, 16, *17*, 35, 37-40, 43, *125*, 129
Chieftain	9-10, 13-16, 43, *74*, *110*
combat skills	41-4
advance to contact	17-20
ambushing	49-52, 60
counter mobility	25-8
counter penetration	34-5
destruction	65-8
mobile defence	33-6
targeting	9-12, 44
Kurassier	*123*
Leopard 2	*10*
M1 Abrams	6, *17*, 20, 29-32, *33*, 36, *122*
M47	5
M48	5
M60	5-8, *18-19*, *34*, 36
M67 flamethrower	*41*
Merkava	21-4

PT-76	*131*
Scimitar	*68*
Sheridan M551	*116*, *130*
T-54 series	61-2
T-55 series	61-4
T-62	62, *65*
Tanque Argentino Mediano (TAM)	96
terrain driving	84
thermal imaging (TI)	90, *91*
track plan	89-90, *91*
trenches	26-7, *28*
turrets	
AMX-10P	158-9
Cascavel	87-8
Dragoon	134-6
Eland	107
Hispano-Suiza HE-60 series	108
Lynx-90	106, *106-7*
M60	7
Marder	94-5
Saxon AT105	143-4
ZSU-23-4	138-40
twin combination, Commando	171-2
Unimog	*98*
United States of America	
AAV7	117-20
Bradley	108, 173-6
Commando	169-72
Dragoon	133-6
M1 Abrams	6, *17*, 20, 29-32, *33*, 36
M47	5
M48	5
M60	5-8, *18-19*, *34*, 36
M67 flamethrowing tank	*41*
M113	*116*, *133*, 165-9
US Marine Corps, AAV7 testing	*118*, *120*
VCI *see* Véhicule de Combat de L'Infantrie	
Véhicule de Combat de L'Infantrie (VCI)	180
Vietnam, APCs	*113-14*, 165-8, 173
West Germany *see* Germany	
wire, counter mobility	26